Foreword

The reports contained in this volume were prepared under the 1991 "Socio-economic Aspects of Climate Change" activity of the OECD Environment Committee. The opinions expressed are those of the author, and do not necessarily reflect the views of individual OECD Member countries.

The book is published under the responsibility of the Secretary-General.

ALSO AVAILABLE

Climate Change: Evaluating the Socio-Economic Impacts (1991)
(97 90 02 1) ISBN 92-64-13462-X   FF130   £16.00   US$28.00   DM50

Environmental Policy. How to Apply Economic Instruments (1991)
(97 91 03 1) ISBN 92-64-13568-5   FF115   £15.00   US$28.00   DM45

Responding to Climate Change. Selected Economic Issues (1991)
(97 91 04 1) ISBN 92-64-13565-0   FF150   £21.00   US$36.00   DM62

Prices charged at the OECD Bookshop.

THE OECD CATALOGUE OF PUBLICATIONS and supplements will be sent free of charge
on request addressed either to OECD Publications Service,
or to the OECD Distributor in your country.

CONVENTION ON CLIMATE CHANGE

Economic Aspects of Negotiations

ORGANISATION FOR ECONOMIC CO-OPERATION AND DEVELOPMENT

ORGANISATION FOR ECONOMIC CO-OPERATION AND DEVELOPMENT

Pursuant to Article 1 of the Convention signed in Paris on 14th December 1960, and which came into force on 30th September 1961, the Organisation for Economic Co-operation and Development (OECD) shall promote policies designed:

— to achieve the highest sustainable economic growth and employment and a rising standard of living in Member countries, while maintaining financial stability, and thus to contribute to the development of the world economy;

— to contribute to sound economic expansion in Member as well as non-member countries in the process of economic development; and

— to contribute to the expansion of world trade on a multilateral, non-discriminatory basis in accordance with international obligations.

The original Member countries of the OECD are Austria, Belgium, Canada, Denmark, France, Germany, Greece, Iceland, Ireland, Italy, Luxembourg, the Netherlands, Norway, Portugal, Spain, Sweden, Switzerland, Turkey, the United Kingdom and the United States. The following countries became Members subsequently through accession at the dates indicated hereafter: Japan (28th April 1964), Finland (28th January 1969), Australia (7th June 1971) and New Zealand (29th May 1973). The Commission of the European Communities takes part in the work of the OECD (Article 13 of the OECD Convention). Yugoslavia has a special status at OECD (agreement of 28th October 1961).

Publié en français sous le titre :

CONVENTION SUR LE CHANGEMENT CLIMATIQUE :
ASPECTS ÉCONOMIQUES DES NÉGOCIATIONS

Table of Contents

Chapter 1

Alternative instruments for negotiating a global warming convention

Chapter 2

Side payments in a global warming convention

Chapter 3

Free-rider deterrence in a global warming convention

Preface

Negotiations leading toward an international Framework Convention on Climate Change are presently underway. These negotiations are taking place under the aegis of the United Nations. If all goes as anticipated, a Convention will be ready for signature at the June 1992 UN Conference on Environment and Development in Brazil.

There are many factors that will enter into the final form of such a Convention. History, culture, science, economics, and a host of other disciplines will each influence the discussions.

Since 1989, the Organisation for Economic Co-operation and Development (OECD) has been actively studying the *economic* dimensions of the global warming debate. This work has dealt with a wide range of economic issues, including: *i)* the *costs* of strategies designed to respond to the problem (whether these involve either abatement or adaptation strategies); *ii)* the *benefits* of responding (i.e. how large the impacts of climate change are likely to be, and which economic sectors will be the most affected); and *iii)* what role *economic incentives* (e.g. taxes, emission permits, or subsidies) might play in the policy response.

A fourth major area of interest for OECD has been the *negotiating process* itself. Being an international organisation, the OECD is vitally interested in how economic variables might influence the nature of any Convention that is eventually negotiated, especially as these variables might affect OECD member-countries. Dr. Scott Barrett, of the London Business School, has been a major contributor to OECD's work in this particular area.

This volume presents three papers by Dr. Barrett that address three different economic aspects of the "negotiation problem". Chapter 1 considers how the choice of a policy instrument might influence which countries would want to join an international coalition on climate change, and under what circumstances they might wish to do so. Chapter 2 looks at the question of how the number of countries participating in an agreement might be increased by the use of so-called "side payments", without sacrificing overall economic efficiency objectives. Chapter 3 then analyses the difficult problem of deterring "free-riding" in an international agreement.

Although the three papers presented here focus largely on *economic efficiency* questions, it is fully recognised that economic efficiency will not be the only, or even necessarily the most important, factor influencing the global warming negotiations. Clearly, other objectives, such as *social equity* and *environmental effectiveness,* will have important roles to play in determining the eventual policy response. Nor is it implied here that economic instruments are always the most appropriate policy options available.

7

Negotiators will, and should, use whatever instruments seem to best fit the particular situation at hand. Economic instruments may be appropriate for certain conditions, but inappropriate for others. In focusing his analysis on economic instruments, Dr. Barrett is simply trying to demonstrate where and how the twin social objectives of economic efficiency and reduced global warming may be partially reconciled in the negotiating process.

All three Chapters take a "simulation" approach, coupled with a "game theory" perspective, to illustrate their conclusions. In no way, therefore, should they be interpreted as indicating the actual negotiating positions likely to be taken by individual OECD governments (or, for that matter, by non-OECD governments). The numbers presented in the text are merely illustrative, as are the general conclusions. Nevertheless, taken together, they provide an overview of the (often counter-balancing) economic forces likely to be at work during the upcoming negotiations.

On the one hand, they suggest that the road to a successful Framework Convention on Climate Change will be long and difficult. On the other, they also suggest areas where progress is most likely to occur, provided the international will is strong enough. Whichever tendency proves to dominate, economic considerations will be near the centre of the upcoming negotiations. This volume offers negotiators some insight into how these economic considerations are likely to influence their important work.

Negotiating a framework convention on climate change: economic considerations

by

Scott Barrett*

* Assistant Professor of Economics, London Business School, and Research Director, Center for Social and Economic Research on the Global Environment (CSERGE), London, U.K.

Chapter 1

Alternative instruments for negotiating a global warming convention

INTRODUCTION AND SUMMARY

It is often assumed that a global warming Framework Convention will be negotiated on the basis of emission reduction obligations, expressed perhaps as a percentage of some base-year's emissions (or, equivalently, as a ceiling on emissions). While that may be so, there are many other instruments that could be used to negotiate a Convention. For example, countries could negotiate energy efficiency measures; reductions in, or the elimination of, subsidies on fossil fuels; economic instruments, such as pollution taxes or tradeable pollution permits; or some combination of these obligations.

There is precedent for many of these possibilities. The Sulphur Emissions Protocol to the Convention on Long-Range Transboundary Air Pollution imposes a percentage abatement obligation on signatories, taking 1980 as the base-year. The Agreement on an International Energy Programme includes provisions for demand restraints. The reduction and removal of subsidies is an important concern of the General Agreement on Tariffs and Trade. The Montreal Protocol on Substances that Deplete the Ozone Layer, especially in its June 1990 revision, includes provisions for international trading of emission allowances.

The specific instrument that serves as the basis for negotiations is important for several reasons. It will have implications for the cost-effectiveness of emission reductions; for the total abatement and total net benefits achieved by a Convention; and for the net benefits of individual countries – both signatories and nonsignatories. This chapter explores the implications of alternative instruments for negotiating a global warming Convention. The paper assumes that side payments are forbidden, except insofar as they may be effected by internationally tradeable emission permits. The chapter also takes only limited account of the free-rider problem, and how this can affect negotiations. These two issues are treated more fully in Chapters 2 and 3.

The purpose of this chapter is *not* to characterise an "optimal" Convention, or to say how such a Convention *should* be constructed. Rather, its purpose is to explore how the choice of a negotiating instrument might affect the final outcome of negotiations. It takes the position that each and every country will negotiate a Convention that is best for *it and it alone.* To be clear, I am not saying that a country *should* have a preference for its own welfare, and no one else's. Nor am I saying that each and every country *does* have preferences for its own welfare, and no one else's. What I want to explore is the question.

if each and every country cared only about its own welfare, and no one else's, how might different instruments affect the outcome of Convention negotiations?

Political theories about the nature of international co-operation suggest that this is a compelling perspective from which to analyse negotiations (see, for example, Keohane, 1984). However, the usefulness of taking this position can perhaps be best seen by considering the opposite position. Suppose each and every country cared equally about the welfare of other countries, so that each country wanted to take decisions that made the whole world as well off as possible. Then negotiations would not be necessary; co-operation would not be a problem. Each and every country would, on its own, take decisions that were in the best interests of all countries; a Convention could do no more than to formalise the unilateral actions of all countries.

Assuming that countries do not care about the well-being of other countries directly is not the same as assuming that the well-being of different countries are independent. Suppose every country would be made worse off by climate change. Then greater world-wide abatement would be in the interests of all countries; actions that improved the well-being of one country would also improve the well-being of all other countries. Interests would be consonant. The problem arises from the fact that if the greater abatement is achieved by the actions of country A, then country A must bear a cost that countries B, C, D, etc. ... do not bear. Countries B, C, D, etc. ... can only gain, since more emissions are abated. However, Country A will lose if the cost of the greater abatement exceeds the associated benefit that *it* receives. The global warming problem is not one in which national interests are incompatible. However, it is one in which national interests are in conflict.

In assuming that each and every country cares only about its own welfare, it may seem that concerns for equity are being ruled out. However, this is not true. What is being ruled out here is the idea that decisions will be taken *for the purpose* of achieving equity. The outcome that is achieved may reflect – indeed, is likely to reflect – concerns for equity, even if the negotiating parties do not care about equity *per se*. This is for two reasons.

First, the outcome that is best for individual countries may also be one that appears to be equitable. Indeed, because an outcome must be agreed and not imposed, to some extent the outcome must be fair. If an offer is unfair in the sense that it makes a country worse off if it agreed to the offer, then the country will not agree, and this (credible) refusal to agree may harm the country or countries that would have benefited had the offer been accepted. Hence, these countries would have some incentive to make a better – i.e. fairer – offer.

Second, outcomes that appear to be equitable may serve as *focal points* for negotiation (Schelling, 1960). Consider a simple example. Suppose two parties must agree on how to divide a pie. Party A would like to have the whole pie, leaving nothing for B. Likewise, B would prefer to have the pie to itself, even if that meant A got no pie. A "rules manipulator" tells both parties that they must write on a piece of paper the share of the pie that they would like to have, and if the shares claimed by both parties sum to less than 100 per cent, then they each will get the share they claimed. If, however, the shares exceed 100 per cent in total, then each gets no pie. What share will the parties claim? The outcome is not certain. However, it is likely that each party will claim a 50 per cent share. Notice that the parties do not split the difference because that is seen to be equitable. They split the difference because an equal division of the pie is the most obvious focal point for this problem.

It may also seem that the perspective taken here rules out any concern for cost-effectiveness. However, this interpretation is also incorrect. It is true that it is assumed here that countries will not *seek* to reach an agreement that is cost-effective. However, one conclusion of this chapter is that the agreement that is best for all of the parties involved will be a cost-effective agreement. Since only economic instruments can guarantee (provided certain assumptions hold) cost-effectiveness, this means that economic instruments are superior instruments in negotiating a global warming Convention.

Many proposals have been made for employing economic instruments in a global warming Convention (see Barrett, 1991*c*). The principal reason is that economic instruments would serve to hold down the costs of an abatement policy. However, if economic instruments are negotiated in a Convention, they will also have *distributional* effects, and hence influence the number of signatories, their "types", and the abatement undertaken.

For example, a Framework Convention which required that signatories impose an identical carbon tax, and retain for themselves the tax revenue, would probably be preferred by countries with high marginal abatement costs (they would undertake less abatement, all else being equal). A programme of internationally-tradeable permits, where permits were allocated initially on the basis of population, would seem to benefit countries with low emissions per capita, all else being constant. (In fact, we shall see later that these results will not always hold because all else is *not* being held constant in these comparisons; other countries are changing their abatement decisions as well, and these changes can influence any given country's preferred choice of instrument.) Hence, different countries would agree to these different proposals. Depending on the distribution of countries according to their "types" (that is, the nature of their abatement costs, GNP, population, etc.) more countries will agree to one proposal than to another, and more abatement will be undertaken under one proposal than under another. The main purpose of this chapter is to examine how the choice of instrument could be expected to influence the outcome of negotiations.

The chapter is divided into two sections. The first describes the bargaining problem. The second provides an analysis of the bargaining problem. The following basic conclusions are reached:

1. An effective Framework Convention would have to include countries with very different characteristics. This means that the bargaining problem is non-trivial.
2. The characteristics that are likely to prove most important to negotiations are each country's base-year emission level, the extent to which marginal abatement costs rise with the level of abatement, the sensitivity of each economy to climate change, and the size (GNP) and population of each economy.
3. Science provides a number of focal points for negotiation – such as a level of global emission which ensures that atmospheric concentrations of greenhouse gases are stabilised. However, none of these focal points seems to be very compelling. Hence, it is unlikely that negotiations will be limited to these.
4. There does not seem to be an obvious focal point for negotiating equal percentage reductions in emissions. Different countries have proposed different levels. What is more, this instrument of uniform percentage abatement may not be acceptable to some countries because of their widely differing cost and benefit functions.

13

5. Economic instruments are superior negotiating instruments to uniform abatement obligations alone. When the former are used, abatement is greater, and each negotiator is made better off. Hence, a global warming Convention should either include provisions for trading abatement obligations, or involve the setting of a uniform carbon tax.

6. Current estimates of the marginal costs of abating CO_2 and the associated marginal benefits suggest that negotiations involving OECD countries are likely to prove most successful if they are based on a uniform emission reduction obligation, with provisions for trading, or on a uniform carbon tax.

7. Agreement between the OECD countries and other important greenhouse gas-emitting countries – including the former USSR, China and Brazil – is likely to prove difficult, if not impossible, if negotiations are limited to choosing a *uniform* obligation, and if side payments are forbidden. This is true even if the uniform obligations are in the form of a carbon tax or tradeable carbon permits. The OECD countries may accept uniform per cent abatement on its own, but the non-OECD countries will probably not. On the other hand, many (but not all) non-OECD countries will want to negotiate emission permits allocated initially on the basis of a ceiling on emissions per capita, but the OECD countries will not. Failure to reach broad agreement seems to be due to three reasons:

   a) Many non-OECD countries will not participate in a Framework Convention unless they are allowed to bear a smaller relative burden. This is partly because the characteristics of these countries – including high emissions per unit of GNP – provide them with a weak incentive to abate emissions in the event that co-operation fails. The OECD countries have a greater incentive to abate emissions if negotiations fail. Hence, if agreement cannot be reached, the non-OECD countries benefit from abatement by the OECD countries without having to undertake much expensive abatement themselves; the non-OECD countries, therefore, have a strong incentive to ''free-ride''.

   b) Another problem is that the instruments considered here serve to polarise the two groups of countries. Negotiations between the OECD and non-OECD countries would prove more successful if an alternative instrument could be found which did not so obviously favour one group over the other. However, a simple instrument that achieves this objective may not exist. What is more, if such an instrument could be found, it may serve to undermine the overall achievement of negotiations – for example, it may win agreement between the OECD and non-OECD, but not prove as successful in intra-OECD negotiations as the blunter instruments.

   c) An alternative, and perhaps more fruitful, approach would be for the non-OECD countries to be allowed to meet weaker obligations, and for the OECD countries to assist these countries by making side payments that compensated the non-OECD countries for the increase in their abatement costs, while not providing these countries with a surplus. The non-OECD countries would like to receive a surplus (such as would occur if the OECD countries were to accept uniform per capita emission ceilings). But negotiations can only succeed if all parties are made better off with an agreement than without it, and the OECD countries will only likely participate in an agreement if such a surplus is small, or non-existent. In this sense, paradoxically, reducing the transfers to non-OECD countries might actually make these countries better off in the long run.

What makes negotiation difficult is that countries are different. If countries were identical, they would each want to choose the same uniform obligation, and agreement would be easy to reach (for a proof, see Barrett, 1991*b*). Agreement may still be difficult to sustain because of the temptation to free-ride; but the initial agreement would not be difficult to reach. Where countries are different, this result could not be expected to hold.

Is broad agreement necessary?

If we assume that all countries must participate in a global warming Convention, then it is obvious that differences among countries cannot be avoided. However, it is not obvious that agreement would have to involve a very large number of countries. If the sources of emissions, or fossil fuel reserves, were highly concentrated, agreement by just a few countries could achieve a great deal. For the bargaining problem, what really matters is whether there exists a small group of relatively homogeneous countries that can affect atmospheric concentrations of greenhouse gases substantially. Unfortunately, there does not appear to exist such a group.

Table 1.1 provides one set of estimates of the main sources of greenhouse gas emissions. In presenting these and other data, the intent is not to imply that they are correct, or that they will necessarily be acceptable to all negotiating parties. Choice of data will be one of the problems that will have to be resolved through negotiations. The data are used here solely as a basis for discussion.

Suppose it were felt that agreement had to be reached among countries responsible for at least two-thirds of emissions (as required by the Montreal Protocol in the case of

Table 1.1. **Greenhouse Gas Emission Shares, 1987**

| | CO₂ Emissions | | CO₂+CH4+CFC Net Emissions[2] |
	Fossil Fuels	Total[1] per cent	
OECD	45%	30%	39%
US	22	15	17
EC	14	9	14
Japan	4	4	4
ex-USSR	18	12	12
Eastern Europe	6	4	3
China	10	10	6
Brazil	1	15	10
India	3	3	4
Indonesia	–	3	2
Mexico	1	1	1
Myanmar	–	2	1

1. Includes emissions resulting from land use changes, including deforestation.
2. Estimates include CO₂, methane, and CFC emissions, weighted by their heat-trapping potential.
Source: Compiled from World Resources Institute (1990), Tables 24.1 and 24.2.

CFCs). Then Table 1.1 indicates that agreement among all the OECD countries would not be sufficient, however emissions were measured. If fossil fuel emissions of CO_2 were controlled, the agreement would probably have to include the former Soviet Union (since that country is responsible for such a huge share of the total) and perhaps some East European countries as well. If all sources of CO_2 (including deforestation) were controlled, the OECD, ex-USSR and east European countries become relatively less important, and the agreement would probably have to include some non-industrialised countries, like Brazil and China. The same would probably be true if methane and CFCs were controlled along with CO_2.

If it were felt that signatories to the Framework Convention had to account for at least three-quarters of emissions, then the number of signatories would have to increase, as would the diversity within the group of signatories. The same would be true if the problem were approached from the supply side – that is, if agreement were to limit the availability of fossil fuels, rather than their use. Two thirds of total coal reserves are found in the United States, the former USSR, and China; three quarters of total oil reserves in the OPEC countries; and three quarters of natural gas reserves in the OPEC countries and the former USSR. Hence, it seems unlikely that an effective agreement can be reached among fairly homogeneous countries.

The problem is even more acute if one considers how the relative importance of these countries is expected to change over time. The OECD countries currently account for 45 per cent of CO_2 fossil fuel emissions, but this share is expected to fall to 37 per cent by 2005 (IEA, 1989)[1]. Any unilateral action by the OECD countries as a whole would not have a profound effect on global concentrations, at least not in the long run. Furthermore, even within the OECD group of countries, substantial differences exist, as is shown later.

How do countries differ?

What characteristics of countries matter for international negotiations on the greenhouse effect? A comprehensive list would be very long indeed. It would include the nature of political representation; the importance of nongovernmental organisations in national decision-making; the country's reputation in complying with past obligations (can the party be trusted?); any linkages that might be thrown up involving other issues (a matter discussed in Chapter 2); and even the acumen and personalities of the negotiators.

At the strategic level, it can also matter whether ratification by a body independent of the negotiator is required (Raiffa, 1982). For example, the executive branch of the United States government has responsibility for negotiating environmental treaties, but the Senate would have to ratify a treaty before it became binding on the United States. The views of these two branches of government can differ markedly, and not all treaties signed by the United States have been ratified. One difficulty with ratification is that the ratifying party can ask for changes or adjustments in the treaty. The "fast-track" provision can help insofar as it prevents Congress from amending the treaty, but it also imposes a deadline, and this may also affect negotiations. It is not clear how ratification would affect negotiations generally, but the nature of ratification is likely to prove important. In some countries, ratification is a formality, because of the nature of political representation, and hence does not really change the bargaining problem. In other countries, ratification requires approval by a different party – one that may hold views that differ from the negotiator's.

There are bound to be different views, even behind the face of a single negotiator (Raiffa, 1982). The UK negotiator, for example, would have to win broad approval among departments representing widely different interests: the Department of the Environment, the Department of Energy, the Department of Trade and Industry, the Treasury, and the Foreign Office. Conflicts of interests are sure to lie behind every negotiator, while the balance among these conflicts will likely vary from country to country. Again, it is not obvious how these internal conflicts can affect negotiations, but they may prove to be important.

Countries are almost certain to have different views about the "just" rule for burden-sharing. For example, some countries might argue that the industrial countries should bear a bigger share of the burden not just because they are richer, but also because they have contributed a much greater share of cumulative emissions, and it is *cumulative* emissions, rather than the current year's emissions that will determine the extent of potential climate change. Other countries might argue that past decisions were taken without any knowledge of these potential changes, and that one cannot undo what has already been done. *Incremental* additions to the concentration of greenhouse gases in the atmosphere will depend on current and future emission levels, and since every country emits these gases, every country must play a role in abatement.

Many of the other important differences can be summarised under the headings, "costs" and "benefits". Costs refer to what must be given up to comply with a Convention obligation, while benefits refer to what is returned as a consequence of having undertaken that obligation. The analysis in this paper is based on these latter differences. The approach taken is to model the objective functions of each of the negotiating countries, and to then examine the best choice for a given country, given the choices of the other countries. Actual negotiations will of course be much richer; they may involve not only the considerations discussed previously, but also strategic behaviour. Strategic considerations are discussed in the Chapter 3 discussion of free-riding.

The remainder of this section discusses the nature of costs and benefits, and of the instruments that could be chosen for negotiation.

Costs

Abatement costs will depend not just on the level of abatement, but also on the policy that is employed to achieve that level. There are, of course, many policies that could be chosen. This chapter assumes that national policies achieve any given level of abatement at minimum total cost or, alternatively, that negotiations will proceed *as if* any given level of abatement by a country would be achieved at minimum total cost. Abatement achieved solely by a moratorium on coal use or a gasoline tax would not be cost-effective because these policies would not ensure that marginal abatement costs were equalised across all sources in an economy[2]. Hence, the possibility of such national policies affecting the negotiations is being explicitly ruled out here. Notice that I am not assuming that *global* abatement is cost-effective, or even that abatement within the group of co-operating countries is cost-effective. (However, it will also turn out that countries will prefer agreements that achieve collective abatement at minimum cost.)

We know quite a lot about the nature of the costs of abating greenhouse gases. We know that if prices reflect social opportunity costs, then the cost of abating the first unit of emission is close to zero. The reason is that, by definition, polluters are indifferent between emitting or abating the last unit of emission. Some analysts have argued that the cost of abating the first unit of emission is *negative* because of insufficient incentives for energy conservation. This may be factually true. However, this Chapter assumes that actions that would increase efficiency will be taken unilaterally (i.e. outside the context of negotiating a global warming Convention). Alternatively, negotiations are assumed to proceed from a baseline where all prices reflect social opportunity costs. (Without this assumption, there may exist opportunities for strategic behaviour; countries may want to take decisions that serve no useful purpose, except to yield an advantage in negotiations.)

We also know that the cost of abating an additional unit of emission must rise as the level of national abatement increases. Empirical evidence suggests that marginal abatement costs increase at an increasing rate, at least beyond some point (see Nordhaus, 1990). This relationship would seem intuitive. One would expect the marginal cost of abatement to rise slowly at first. However, as abatement becomes nearly complete, marginal costs must become very large. Indeed, given current technology, one would expect the cost of abating the very last unit of emission to be huge. There is nothing in principle that would prevent us from specifying a non-linear marginal abatement cost schedule in the modelling exercise. However, to simplify matters, I shall assume that the marginal abatement cost schedule is linear. Not much will be lost in making this assumption, provided the schedule is approximately linear over the range of abatement levels that might actually be negotiated[3].

Marginal abatement costs will vary among countries for several reasons. One is that different countries will have different substitution possibilities. Another is that countries will have different base-year emission levels. Depending on how abatement is defined, marginal abatement costs may also vary because of differences in the rate of economic growth.

If two countries have identical emissions in the base-year, but one has greater opportunities for substituting out of polluting activities, then this country would have a lower marginal abatement cost over all positive abatement levels (all else being equal). To take an example, Norway and Finland emit about the same quantity of CO_2. However, Finland burns coal and natural gas as well as oil (and produces electricity from nuclear power). All else being equal, one would expect that the marginal cost of abatement would rise more steeply for Norway than Denmark because Norway burns almost no coal, and does not have a gas grid (Norway also has no nuclear stations). Norway's emissions cannot be reduced by substituting away from coal, and substitution of gas would require costly investment in transmission and distribution.

If two countries are identical in their substitution possibilities, but have different base year emission levels, then one would expect that the marginal abatement cost would be identical for any given *percentage* abatement, but that the larger country – the country with a higher base emission level – would have a lower marginal abatement cost for any *absolute* level of abatement. To take an example, the United States and United Kingdom have similar substitution possibilities insofar as the consumption shares of coal, oil, gas and non-fossil fuels are nearly identical in the two countries. However, CO_2 emissions are nearly eight times greater in the US than in the UK. One would expect that a 10 per cent reduction in emissions would entail a similar marginal abatement cost for the two countries, because they each have about the same opportunities for substitution. However, total abatement costs would naturally be higher in the US, because the US would be abating a much greater quantity of emissions. If the two countries abated emissions by the same amount in absolute terms, then the percentage abatement would have to be much greater for the UK, and as a consequence marginal abatement costs would have to be much greater for the UK than the US.

Pulling these considerations together suggests that marginal abatement costs for country i may be represented by the equation:

$$MC_i(q_i) = c_i q_i / e_i, \tag{1}$$

where q_i is the absolute level of abatement undertaken by country i, e_i is base-year emissions, q_i/e_i is percentage abatement, and c_i is a technical parameter which reflects the ease with which i can abate its emissions.

Integrating [1], and using the assumption that the cost of abating one unit of emission is close to zero, yields the total abatement cost function:

$$TC_i(q_i) = c_i q^2_i / 2e_i \tag{2}$$

For any given country, c_i and e_i are assumed to be given. As noted earlier, the base year and the baseline from which emission reductions are to be calculated must themselves be negotiated. However, it is assumed here that the base-year, the baseline and all relevant data have already been agreed. The e_i are, therefore, known to all negotiating parties. The baseline assumes that all emissions that can be abated at a cost of zero, or less, have been abated. Hence, the marginal cost of abating the first unit of emission is close to zero. Abatement by any country is assumed to be cost-effective, and this assumption defines c_i. Of course, the abatement cost functions are subject to some uncertainty. However, it is assumed here that the c_i parameters are public knowledge. What must be negotiated, of course, is q_i or an instrument which determines q_i. Because c_i and e_i determine costs, they will also help determine the negotiated outcome.

It may seem surprising that the cost function was derived without making any distinction between low- and high-income countries. However, what is special about the low-income countries is that their economies are expected to grow at a faster rate than those of the high-income countries. We should therefore also expect that their emissions in the absence of abatement will grow at a faster rate[4]. Faster growth by the low-income countries is likely to prove particularly important to negotiations if what is being negotiated is a fixed obligation – like a ceiling on emissions.

To see this, consider the hypothetical data presented in Table 1.2. A low-income and a high-income country are both committed to meet the Toronto conference targets; each has agreed to ensure that its emissions do not exceed 80 per cent of its 1988 level from 2005 onwards. The countries have identical base-year emission levels (100 in 1988). However, the low-income country's emissions are expected to grow 2 per cent annually in the absence of abatement, while the high-income country's emissions are expected to grow 1 per cent per year. The per cent abatement required to meet the target is greater for the low-income country, because this country is expected to grow faster. It is primarily for this reason that total abatement costs would be greater in low-income countries, all else being equal.

What distinguishes the high- and low-income countries, then, is not the fact that they have different cost parameters, but that the burden of a given emission ceiling obligation will be greater for the low-income countries, all else being equal, because they are

Year	Emissions in Absence of Abatement	Emission Ceiling	Abatement	Per cent Abatement
(1)	(2)	(3)	(2)-(3)	(4)/(2)
			(4)	(5)
High-Income Country (Emissions in Absence of Abatement Grow 1% per year)				
1988	100	–	–	–
2005	119	80	39	33%
2050	186	80	106	57%
Low-Income Country (Emissions in Absence of Abatement Grow 2% per year)				
1988	100	–	–	–
2005	140	80	60	43%
2050	346	80	266	77%

expected to grow more rapidly. Given the form of equation [1], however, it makes no difference whether this greater burden is reflected in a higher relative obligation (q_i/e_i) or in a higher cost parameter (c_i).

Benefits

Abatement benefits are defined as the minimum of damage plus adaptation costs avoided by greater abatement. One should not calculate benefits as the damage that would be caused if no adaptation were undertaken, because adaptation can lower damage costs. For example, if agricultural losses can be cheaply reduced by switching crops, then these adjustments should be reflected in abatement benefits. If damage in the absence of adaptation and abatement were, say, $10 million, but that damage would be reduced to $2 million if $5 million were spent on adaptation, then abatement which resulted in no climate change (and, hence, no damage) would provide a benefit of $2 plus $5 or $7 million.

Not much is known about the marginal abatement benefit schedule, apart from the fact that marginal abatement benefits for country i depend on *global* abatement (or, alternately, on global emissions). It is believed that many countries will be harmed by climate change, but that some may possibly benefit, at least for small changes in climate regimes.

In modelling benefits, it would seem sensible to assume that marginal abatement benefits are zero for some level of abatement. For example, we might assume that marginal abatement benefits are zero when all emissions are abated. This would make sense for normal pollution problems. If there were no lead in the air, then the damage associated with adding one small unit of lead would be near zero. Higher concentrations of lead would be associated with higher and, likely, increasing marginal benefits of abatement. But once one accepts that climate will change even in the absence of pollution, or that some changes will actually be beneficial for some countries, it is not obvious that marginal abatement benefits will equal zero when emissions are zero – at least not for every country; they may be positive for some countries, and negative for others.

Another difficulty is determining the shape of the marginal abatement benefit schedule. For a country that is expected to be harmed by global warming, it would seem reasonable to suppose that marginal damage will increase with the level of warming. However, even if this is accepted, it may not follow that marginal abatement benefits always decline with the level of *abatement.* Because of the many feedbacks involving the oceans, clouds, and ice caps, it is conceivable that the marginal abatement benefit schedule will not be monotonic over the entire range of feasible abatement levels. For the purpose of the present analysis, it is assumed that marginal abatement benefits are not only monotonic, but *linear* in the level of global abatement. However, it must be remembered that this assumption is particularly questionable[5].

Abatement benefits will also be determined by how sensitive an economy is to climate change, and by the absolute size of the economy. Some countries, by virtue of their location and economic structure, will be especially vulnerable to climate change. Bangladesh is an example. It is low-lying, and hence is sensitive to changes in sea level. Its economy also relies heavily on agriculture (agriculture's share of GDP is 46 per cent in Bangladesh, but only 2 per cent in the United States; see World Bank, 1990), and this sector may be especially harmed by climate change in marginal areas[6]. However, the total damage which an economy could suffer in any year must be (approximately) bounded from above by GDP[7]. Bangladesh may be relatively more vulnerable to climate change than the United States, all else being constant, but the *total* damage caused to the United States could be much greater for any level of warming because the US economy is 25 times greater in size (World Bank, 1990).

Discussions about a global warming Convention invariably focus on the differences between rich and poor countries. To some extent, these discussions reflect ethical views about how the burden *should* be distributed among these countries. However, the bargaining problem will also be influenced by other differences. One is the obvious fact that poorer countries, by virtue of their lower incomes, have less to lose from global warming than rich countries, all else being equal. Another is that poorer countries would probably discount the future at a higher rate than richer countries, and hence would attach a smaller value to abatement benefits, all else being equal, since these benefits will be deferred for many decades. Both of these considerations would serve to make the marginal abatement benefit curve lower for poorer countries, holding all else constant.

This discussion suggests that the marginal abatement benefit function for country i could be written:

$$MB_i\,(Q) = b_iG_i(a_i - Q/E),\hspace{3cm}[3]$$

where Q is global abatement; E is global emissions in the absence of any abatement; a_i is a parameter determining the intercept of the marginal abatement benefit function; b_i is a parameter reflecting both the sensitivity of an economy to climate change and the weight which is attached to deferred benefits; and G_i is i's gross domestic product. Notice that, if $Q = E$ implies $MB_i = 0$, then $a_i = 1$. For reasons discussed above, a_i may be greater or less than 1. However, the simulations carried out later in the paper all assume that $a_i = 1$ for all i.

Integrating [3] gives the total abatement benefit function:

$$TB_i(Q) = b_iG_i(a_iQ - Q^2/2E) \qquad [4]$$

Again, b_i, G_i, a_i and E are given, but will affect a country's choice of q_i insofar as they affect TB_i. A country's choice of q_i will also be influenced by the abatement choices of other countries, since these influence Q. It is this fact which gives rise to the free-rider problem.

Notice that if each country cares about its own net benefits and not those of any other country, then this is all the information we need to determine whether or not an agreement can be reached[8]. If a global planner were to solve for the abatement levels that maximise global welfare, then we would also require information on the marginal utilities of consumption. The global planner might also attach higher weights to the countries that would be harmed by climate change than to those that would benefit. However, for the bargaining problems, these values are not needed. To solve the bargaining problem all we need to know is that a country prefers more net benefits to fewer. We do not need to compare utilities. In the simulation analyses, some results of the total net benefits associated with any given agreement are presented. However, these are calculated without taking account of the ethical issues noted here. Hence, one must be careful in interpreting these estimates. Outcome A may have higher total net benefits than B, but if poorer countries are made better off under B than A, then a global planner may prefer outcome B to outcome A.

Are there obvious focal points for negotiation?

It would perhaps seem most natural to negotiate a global warming Convention on the basis of science. Very often science does provide focal points by showing that environmental damage is insignificant unless concentrations of some pollutant exceed some specified level; this level then becomes the natural one around which parties can negotiate.

Table 1.3 lists a number of possible science-based targets. Careful study of these shows why negotiation based on science alone will prove difficult.

If emissions are reduced in order to stabilise concentrations at current levels, concentrations will still exceed pre-industrial levels; there may still be some climate change.

Table 1.3. **Possible Science-Based Targets for Control**

Target	Rationale
60% reduction in long-lived gases, e.g. CO_2.	Stabilise atmospheric concentrations of greenhouse gases at current levels.
Do not allow global mean temperature to rise more than 2°C above current level.	This level is the maximum temperature experienced during the last several million years in which human beings have existed.
Do not allow the rate of warming to exceed 0.1°C per decade.	This is the rate at which forests and other ecosystems are estimated to adapt to temperature changes without severe dislocation.

Sources: IPCC (1990) and Pearce (1991).

But if we allow this amount of climate change, why not more? This question is particularly pertinent because the *cost* of achieving this particular target would probably be very (unacceptably?) high. Hence, targets aimed at stabilising concentrations are unlikely to prove compelling focal points.

If emissions are reduced such that global mean temperature does not exceed its previous maximal value over the course of human history, the cost will again be high, and the question might be asked, "just because we have not experienced a higher temperature in the past, does it follow that we could not survive (if not prosper) at a higher temperature?" People raising this last question might point out that populations have, over time, moved toward warmer areas because technology has historically made these areas much more attractive to live in (Schelling, 1983).

The final proposed target poses similar difficulties. Surely, some ecosystems will still be vulnerable at a 0.1° C per decade warming. For example, if sea levels were to rise, coastal ecosystems like mangroves would be harmed, if not ruined, and if these ecosystems are going to be lost, people may well ask why others could not also be sacrificed.

There are still other problems with these science-based targets. One is that the last two targets are expressed in terms of temperature and temperature change, whereas it is emissions that would have to be controlled. Hence, the last two targets would have to be translated into emission levels. The relationships between emissions and concentrations, and concentrations and temperature are uncertain (see Pearce, 1991), and this uncertainty will further weaken the basis for negotiating around such targets.

What is more, all these targets were derived from a concern about the *effects* of climate change, and hence can only be achieved through *global* control. Yet, when one considers how this would be accomplished, the task of negotiating these targets seems extremely difficult, if not impossible. Suppose a global emission ceiling were negotiated among all (or nearly all) countries. If one country pulled out of the agreement, the others would have to abate more of their emissions if the global ceiling were still to be met. But then this defector would have saved its abatement costs and lost nothing in abatement benefits. The same would be true of all other defectors. All countries would be better off outside the agreement. And so, the agreement would not be reached in the first place. Hence, it seems that negotiations will probably have to rely on focal points derived from something other than science.

A number of proposals have already been made to simply reduce CO_2 emissions by a given percentage, where the level of reduction has been based on something other than science. Table 1.4 lists several of these possibilities. It should be noted that most countries have not proposed a target for control, including the biggest emitters (the United States, the former Soviet Union, Brazil and China)[9].

The great variety of proposed targets suggests that there is no obvious focal point for negotiating a Framework Convention on the basis of emission reductions either. The difficulty seems not to lie in choosing the base-year. The Toronto Conference, which met in 1988, chose 1988 as the base-year, but all recent proposals have chosen 1990 as the base-year. However, the proposals differ both in terms of their per cent reduction relative to this base-year, *and* the year in which this target is to be achieved[10].

Notice that these proposed targets say what should be done by a particular date, but not what should be done after that date has passed. In this sense, these proposals are similar to the Nitrogen Oxides Protocol, which requires that emissions be reduced 30 per cent from their 1987 level by 1994. This approach may have some virtue, because, by the

Table 1.4. **Proposed Targets for Carbon Dioxide Control**

Proposer	Proposal		
	Per cent Reduction	Relative to Base Year	Achieved by End Year
Toronto Conference	20	1988	2005
2000 Club [1]	0	1990	2000
UK	0	1990	2005
Netherlands	5	1990	2000
	8+	1990	2005
Japan	−5	1990	2000
New Zealand, Italy, Australia, Luxembourg	20	1990	2005
Germany:			
Pre-unification	25	1990	2005
Unified	30	1990	2005

1. EC, Switzerland, Austria, Norway, Italy, Denmark, France, Iceland, Luxembourg, Sweden, Canada. See Pearce (1991) for comments.
Source: Pearce (1991).

time the end-date has been reached, we should have learned more about the problem and the opportunities for dealing with it. But, in setting only a temporary target, some uncertainty is created that could pose problems.

Suppose, for example, that countries agree to stabilise emissions at their current levels by the year 2005. If a signatory wanted to comply with its obligation by using tradeable permits, it could choose annual emission levels that ensured the target was met[11], issue a quantity of emission permits equal to these annual emission levels, and allow these permits to be traded. However, if it is not known what the country's obligations will be in 2006, 2007, and so on, then firms will be uncertain about the quantity of permits that will be available in these years. Experience with the U.S. emissions trading program suggests that firms may be reluctant to trade when they fear being caught short of permits at a later date (Hahn, 1989). The net effect of this uncertainty might be that the trading system would not work properly, even in the years prior to 2006.

A final observation is that emission ceilings would be an inefficient means of achieving any overall level of abatement. Costs would be minimised if the marginal costs of abatement were equal for all signatories. One could, in theory, calculate each country's marginal abatement cost schedule and then calculate shares of a total abatement level which would ensure that marginal abatement costs were equal for all countries. However, it seems unlikely that such elaborate calculations would be attempted. Certainly if one looks for precedents, one finds that obligations are typically uniform (Barrett, 1991*b*).

Instruments for Negotiation

The obvious "focal point" instrument for negotiation, and the one suggested by the proposed targets in Table 1.4, is uniform per cent abatement. However, as just noted, this instrument would not achieve the total level of agreed abatement at minimum cost. A modified instrument would allocate permits to emit CO_2 on the basis of uniform per cent abatement, but then allow inter-country trading of permits. It might seem obvious that this modified instrument would be (at least weakly) preferred by all parties to the simple uniform per cent abatement obligation, for trading would ensure, given the initial burden, that at least one country was made better off, and none worse off. However, if countries choose a uniform abatement obligation *knowing* that trading can and will take place after permits have been allocated, then this knowledge might alter the level of obligation that might be acceptable to a given country. There is at least the potential that one country would be made worse off. However, this seems unlikely, and in all simulations presented below, uniform per cent abatement with trading is strictly preferred by all countries to uniform per cent abatement without trading. Notice that this is precisely the approach taken by the Montreal Protocol, and its subsequent revision in June 1990[12].

An alternative economic instrument to tradeable permits is the emission tax. Carbon taxes have already been imposed in several countries (including Finland, the Netherlands, and Sweden). They have also been proposed in many others, and are now being examined by the European Commission. In the context of a global warming Convention , there are three ways in which emission taxes might be used. The grandest scheme would have an international agency impose an international carbon tax, collect the revenues, and then redistribute these according to some previously agreed formula. The more modest scheme would simply involve countries imposing such a tax unilaterally, as a means of complying with an agreed emission reduction obligation. The latter scheme is not directly relevant to international negotiations. The former is, but also raises the issue of side payments, which is discussed in Chapter 2. An alternative proposal would be for individual countries to negotiate the level of an international emission tax, but allow the tax to be collected nationally. This is the proposal examined here.

The principal advantage of the internationally-negotiated (but nationally-collected) emission tax is that it would ensure that abatement among signatories would be achieved efficiently. In lowering costs, there is at least the prospect that more could therefore be achieved by such a tax. Tradeable permits have this same feature.

However, the two will not be equal in their effects. In fact, some countries may well prefer uniform per cent abatement without trading to an internationally-negotiated carbon tax. To see this, suppose that a tax is imposed at a level which ensures that the same total reduction in emissions is achieved. The tax would be cost-effective; it would achieve the desired reduction at minimum total cost. However, it might raise the cost to some countries, even while lowering costs in total.

Table 1.5 illustrates these points. Country 1 has lower abatement costs than Country 2. If an emission ceiling of 10 units each is negotiated, Country 1 bears a cost of $50, and Country 2 of $100. If these countries are allowed to trade these permits, and trading is perfect, then costs for both countries will fall. Country 1 undertakes more abatement, but is compensated by Country 2 such that its abatement costs fall from $50 to $44. Although Country 2 pays Country 1 to undertake greater abatement, Country 2 no longer has to suffer its own high marginal abatement costs, and is able to lower its total bill from $100 to $89. Total costs are the same if an emission tax is set to achieve the same 20 units of abatement, but the tax demands that Country 1 bear a cost of $89 and Country 2 of $44. Compared with the emission ceiling, Country 2 is better off with the tax, but Country 1 is worse off.

Table 1.5. **Hypothetical Example of Payoffs Associated with Different Instruments**[1]

	Negotiated Emission Ceilings		Emission Permit Trading		Emission Tax	
	Abatement	Costs	Abatement	Costs[2]	Abatement	Costs
Country 1	10	$50	13.33	$44.45	13.33	$88.84
Country 2	10	100	6.67	88.88	6.67	44.49
	20	$150	20.00	$133.33	20.00	$133.33

1. Costs for Country 1 are assumed to equal $q_1{}^2/2$, and costs for Country 2 are assumed to equal $q_2{}^2$, where q_1 and q_2 are the abatement levels for countries 1 and 2, respectively. Abatement benefits for each country are the same in each case, because *total* abatement is held constant.
2. Costs are net of the value of permits that are traded. It is assumed that permits trade at a price equal to the marginal abatement cost of each country, when marginal abatement costs are equal for both countries, and total abatement equals 20 (i.e. $13.33 per permit).

This does not mean that tradeable permits are generally superior to a carbon tax in negotiations. The hypothetical example in Table 1.5 assumes that total abatement is constant in all three cases. However, in actual negotiations the level of abatement acceptable to all parties will vary, depending on the instrument that is used. What will matter in actual negotiations is net benefits, not just costs alone. (This will become clearer when the simulations are examined in more detail later.) Furthermore, although the simulation analysis assumes that trading is perfect in the sense that the marginal abatement costs are equalised across all countries, there is no guarantee that trading will be perfect; there may exist transactions costs or market power problems (see Barrett, 1991c).

However, even if the assumptions of perfect competition *do* hold, the success of the Convention in terms of the total abatement achieved, or the global net benefits gained, will depend crucially on how the permits are allocated initially. It would seem reasonable to suppose that each potential signatory will only agree to a Convention that leaves this country better off. While a country will be better off with trading than without trading, given the same allocation of emission reduction obligations, it is not obvious that a country will be better off with these given obligations, even if trading is perfect.

Permit allocation rules

As just noted, the acceptability of any given tradeable permit proposal will depend on the overall level of control, and how that level is allocated among the signatories. There are many possible allocation rules that could be employed. Table 1.6 presents a number of proposed rules for allocating internationally tradeable emission permits among nations. Other rules could be constructed, but the Table provides a fairly complete listing of the rules that have been discussed thus far.

The first two rules involve countries negotiating a *global* limit on emissions, and then allocating these among nations on the basis either of CO_2 per capita or CO_2 per unit of GNP. One difficulty with these proposals is the very notion that a global limit on emissions and a formula for allocating emission permits could be decoupled. There is no international precedent for such an allocation. More importantly, such a proposal invites

Table 1.6. **Some Proposed Allocation Rules**

Rule	Rationale
1. Negotiate a global limit on CO_2 emissions, and allocate this limit on the basis of CO_2 emissions per capita.	"The moral principle is simple, namely that every human being has an equal right to use the atmospheric resource. The economic principle follows directly – those who exceed their entitlement should pay for doing so. The practical effect is obvious: it would require the industrialised world, with high per capita energy consumption, to assist the developing world with efficient technology and technical services" Grubb (1989, p. 37). See also Pearce (1990), Bertram *et al.* (1989), Hibiki *et al.* (1989), and Burtraw and Toman (1991).
2. Negotiate a global limit on CO_2 emissions, and allocate limit on basis of CO_2 emissions per unit of GNP.	"... carbon emissions are tied to economic activity and the objective should be to maximise the efficiency of economic production" (Grubb, 1990, p. 36). Grubb argues that measurement of GNP would pose a problem, although Pearce (1990) maintains that countries already agree on the basis of GNP measures through the UN statistical system. Both Grubb and Pearce argue that this rule would discriminate against poor countries. Grubb (1990, p. 36) refers to the rule as "an untenable proposition".
3. Uniform percentage reduction in emissions.	Minimal disruption to *status quo ante,* and hence, involves smaller transfers. Common rule used in other international environmental agreements. Criticised by Grubb (1990, p. 36) on the basis that it would "reward the countries which are currently the most polluting."
4. Allocate permits to poor countries.	"Since the world's rich are the chief polluters of the atmosphere, there are strong equity (and some efficiency) grounds for allocating entitlements initially to the world's poor, so that the necessary purchase of entitlements by the world's polluters would generate a financial flow from rich to poor, hopefully providing resources to encourage development of the poor countries." (Bertram *et al.* 1989, p. 14).
5. Allocate permits in inverse proportion to per capita consumption of fossil fuels.	"This would directly reward those countries which moved seriously to renewable energy, while at the same time helping countries with low levels of development, and hence low total energy consumption" (Bertram *et al.* 1989, p. 15).
6. Choose an overall level of cost to be borne by all countries, and then set each country's emission reduction so that abatement cost, relative to pre-control income is the same for all countries.	Based on "Ability to Pay" Principle (Burtraw and Toman, 1991).
7. Choose an overall level of cost to be borne by all countries, and allocate this cost such that abatement cost relative to pre-control emissions is the same for all countries.	Based on "Polluter Pays" Principle (Burtraw and Toman, 1991).

free-riding. If a total reduction is agreed, and a country pulls out of the agreement, then the remaining countries must make up for this withdrawal. The defector saves costs, but suffers no loss in benefits. All incentives favour withdrawal. But then there can be no agreement. The only way such a scheme could work would be if some other form of punishment could be found to deter free-riding.

An alternative approach would simply be for countries to negotiate individual emission ceilings – perhaps on the basis of CO_2 per capita or per unit of GNP. It is actually not necessary that a global ceiling be negotiated. With country-level ceilings, withdrawal from the agreement by one country would not place a burden on the remaining signatories to increase abatement. The remaining signatories could, of course, decrease abatement to punish the defecting country. Indeed, as discussed in Chapter 3, such a punishment may well be necessary to deter free-riding.

However, even if obligations are negotiated at the individual country level, the two allocation mechanisms will have difficulty winning broad acceptance because they each make one group of countries better off and another worse off. As we shall soon see, CO_2 emissions per capita tend to be much higher in the OECD and eastern bloc countries than in the non-industrialised countries. Such a proposal therefore implies large transfers of funds from the industrialised nations to the non-industrialised nations. On the other hand, CO_2 emissions per unit of GNP tend to be low in the industrialised countries, and high in the non-industrialised countries. The proposal to allocate emissions on the basis of CO_2 per unit of GNP will therefore imply transfers from poor countries to rich. The extent of the difficulties that this may cause will be illustrated later.

A simple rule of "grandfathering" emissions by requiring that every country reduce emissions *relative to a base-year* would probably involve smaller transfers, but would be seen to harm poorer countries. One problem with this rule is that it would be seen to be unfair by the poor countries because they are growing more quickly, and hence, for them, the burden would be greater over time, all else being equal. The poor countries may well feel that they should be allowed some room to increase emissions as their economies expand.

Importantly, it is not necessary that countries negotiate on one rule only. For example, the Montreal Protocol demands that all signatories reduce their consumption and production of listed substances from their 1986 levels by 50 per cent by 2000, except for "developing" countries. These countries are allowed an additional ten years to comply, provided their consumption does not exceed 0.3 kilograms per capita. The base level from which consumption must fall 50 per cent will be either the average consumption level over the period 1995-1997 or 0.3 kilograms per capita, whichever is lower. This agreement essentially "grandfathers" abatement by the industrialised countries, but imposes a ceiling on developing countries that is independent of past consumption levels.

In the Montreal Protocol, it happens that any given country must fall into one of two groups: developing or non-developing. However, one could increase the number of groups. The problem of deciding on groupings of countries is discussed in the next section.

Allocation Rules 4 and 5 seek to achieve the same equity objective as the "per capita" rule, but are not well specified. It is not made clear *how* entitlements would be allocated in either case, and it seems unlikely that these rules would serve as practical alternatives in negotiations.

The final two rules are interesting because they are based on economic variables. However, both suffer from the same problem as Rules 1 and 2, insofar as they both require that a *global* cost burden be negotiated independently of the shares to be borne by individual countries. However, as in the case of Rules 1 and 2, the essential features of these rules might be retained while disregarding the global cost constraint. Countries could agree to spend a certain proportion of their GNP on abatement, or an amount which was proportional to pre-control emissions.

The great difficulty with Rules 6 and 7 is that it would be extremely difficult to calculate abatement costs *ex poste,* and hence to monitor compliance. Costs do not depend solely on the amount of money spent on abatement equipment, but on all the actions undertaken to meet an abatement target. To reduce CO_2 emissions, countries would substitute non-fossil fuel energy sources for fossil fuels, less polluting fossil fuels for the more polluting fuels, and capital and labour for energy. Firms would alter their production processes and even the nature of their products. And so on. A review of the literature will demonstrate how difficult it is to estimate the costs of environmental protection for an individual industry, let alone for an entire economy (see, for example, Gollop and Roberts, 1983; and Jorgenson and Wilcoxon, 1989).

Another difficulty with these last two rules is that they tend to not be consistent with the pattern of burden-sharing observed in other situations. For example, it has been shown that countries with a larger GNP tend to bear a disproportionate share of the of the burden of common defence provided by the NATO alliance (Olson and Zeckhauser, 1966)[13].

BARGAINING ANALYSIS

If obligations are allowed to vary from country to country, and are not guided by any particular rule, one might expect that broad agreement could be reached. But one might also expect that negotiations would degenerate to a level where every country agreed to do roughly what it would have done without an agreement. This does not mean that countries would undertake no abatement. As demonstrated elsewhere, countries are likely to face some incentive to take some action unilaterally (see Barrett, 1991a). However, it does mean that co-operation would fail to win any significant incremental action.

Alternatively, if obligations were set according to one fixed rule – such as CO_2 per capita or per cent abatement relative to some base-year – it is likely that a few countries would agree to undertake more abatement than they would otherwise have done. However, it is also likely that most countries would be made worse off by the agreement, and hence would not sign the Convention. Yet, if a substantial number of countries do not sign, then abatement by just a few countries may not affect global abatement materially.

This discussion suggests that there may exist a trade-off between winning broad agreement and negotiating a fixed rule that commits signatories to do more than they would have done had no agreement been reached. One might want to increase the number of rules that could be used in negotiations, for that would win wider agreement. But beyond some point, additional rules will mean that countries undertake less incremental abatement. The challenge is to have enough rules that a substantial number of countries participate in the Convention, but not so many that the participants undertake little incremental abatement.

Suppose one were to negotiate two rules, instead of just one. One rule would apply to one group of countries, and the other rule to another group. Since countries have no difficulty agreeing to obligations when they are identical, it seems natural that countries would organise themselves into groups, where the countries in each group were fairly homogeneous. As an example, the Montreal Protocol makes a distinction between developed and developing countries. Developed countries consume higher quantities of CFCs per capita. Developing countries consume less per capita, but demand in these countries is expected to grow rapidly. The agreement requires developed countries to reduce emissions from their already high levels. Developing countries are allowed to increase emissions for a time, but not above a fixed ceiling.

This Chapter attempts to determine whether a global warming Framework Convention might also be negotiated by two distinct groups. One "natural" group would consist of the OECD countries. The other would consist of non-OECD, large CO_2 emission sources. The analysis first considers the nature of negotiations within the OECD. It is assumed here that agreement by these countries is needed if other countries are eventually to participate in further negotiations. In the next stage, an examination is presented of the nature of negotiations between the OECD as a group, and the other big emitters – the former USSR, China and, if negotiations are to include the effect of deforestation on emissions, Brazil. This grouping is somewhat arbitrary, and others might be chosen. For example, Manne and Richels' (1990) economic analysis divides the world into two groups – one including the OECD and the former USSR and Eastern Europe, and the other including China and all other countries. Whalley and Wigle (1989) divide the world into three groups – the "developed world," the "developing world," and "oil exporters".

Analytical approach

A number of studies have attempted to show how any given instrument would affect individual countries. These range from simple comparisons (see, for example, Pearce, 1990) to sophisticated modelling exercises (see, for example, Whalley and Wigle, 1989 and 1990; and Manne and Richels, 1990*b*). A feature common to all these studies is the complete absence of behavioural reactions. An abatement obligation, or a carbon tax, is *imposed* on the countries. However, one would expect that in real negotiations, the obligation that is ultimately selected will be endogenous and will depend on the circumstances underlying negotiations – the number of negotiators, the characteristics of these negotiators, and the instruments that are employed in negotiations.

In the analysis that follows, it is assumed that, in the absence of successful negotiations, countries will choose abatement levels where the marginal abatement cost of each country is equal to its marginal abatement benefit (this defines the Nash equilibrium in the game that is played). This provides the *status quo* point for negotiations. All countries will agree to a negotiation proposal if all do at least as well as at the *status quo* point (I am not considering the formation of coalitions here, but for some discussion of this possibility, see Chapter 3).

Rather than calculate the solution to the bargaining problem (using a solution concept such as the Nash bargaining solution), the analysis allows each negotiator to choose the level of any given instrument that maximises its own net benefits, under the assumption that the other negotiating parties will comply with the chosen level. An instrument can be acceptable to a country only if the net benefits which the country receives when the instrument is set at the country's preferred level is at least as great as the net benefits which that country receives at the *status quo* point. In turn, an instrument can be effective in international negotiations involving all parties only if it is acceptable to each party.

Depending on the circumstances, none of the considered instruments may be acceptable to all parties – in which case, negotiations involving all the parties are doomed to fail (provided better instruments cannot be found, or side payments are made). However, it is also possible that more than one instrument will be acceptable to all parties. If the latter is the case, then the parties may have different views about which instrument is best. In this case, negotiations must choose an instrument, as well as a level for that instrument (for example, the instrument of a carbon tax plus a uniform tax rate). A country's preferred instrument is taken to be the one which delivers the biggest net benefit to that country (at a level equal to the country's preferred choice), while also being acceptable to the other parties. The analysis does not determine which instrument would be chosen in such cases, or the level at which the instrument would be chosen. To solve for these problems, one would need to employ a specific bargaining solution concept.

In the simulations, values for the parameters in equations [1]-[4] must be used. The analysis employs actual values for e_i, G_i and population; assumes $a_i = 1$ for all i; and substitutes various values for b_i and c_i (or, rather, b_i/c_i, since it is the latter that matters).

One cannot hope to provide quantitative results on the bargaining problem, or even on the bargaining range. The data needed to obtain such results do not exist. There is much uncertainty about the costs of abating emissions, and the science of global climate change is not yet able to determine regional impacts with any precision. However, it may be possible to obtain some qualitative results. For example, it might be possible to show that for a wide range of values for the key parameters in the net benefit functions – b_i and c_i – agreement within one group of countries might be easier to reach using one instrument than another. Such results would not suggest what countries could agree to, but they would provide insights into the nature of the bargaining problem. For example, they may serve the purpose of getting negotiators to ask questions about the nature of the net benefits, and about the types of instruments that could be employed in negotiations. That, of course, is the purpose of this analysis.

Hypothetical OECD negotiations

Nitze (1990, p. 37) argues that agreement among OECD countries is a precondition to wider agreement:

"A commitment by the OECD countries to stabilise their CO_2 emissions is critical to the success of the Convention because the developing countries are unlikely to accept either a global greenhouse gas stabilisation target or separate sub-targets applicable to themselves, unless the OECD countries demonstrate their own commitment to take the greenhouse problem seriously. Representatives of Brazil, India and other developing countries have repeatedly stated at IPCC meetings that their governments will not agree to limit their countries' emissions unless the industrialised countries first commit themselves to reducing their own CO_2 emissions."

Table 1.7 shows that substantial differences in emissions per capita and per unit of GNP exist even within the OECD. The United States is clearly the biggest source of CO_2 emissions. However, U.S. emissions account for less than half of total OECD emissions. Other countries should clearly be brought into the analysis, although most emit only a small fraction of the total.

The problem can be simplified if we treat the European Community countries as a single negotiating body, and examine the potential for reaching agreement between the EC, the United States and Japan. This group of countries makes up 89 per cent of OECD emissions, and it may be that, if agreement can be reached by these countries, then it can also be reached by the other OECD countries.

To see this, notice from Table 1.7 that emissions data tend to be bounded by the United States and Japan. Outside the EC, no country has a higher level of CO_2 emissions per capita than the US, and only Turkey, Switzerland and Austria have a lower level than Japan. Among all non-EC countries, only Turkey, Canada and Australia have higher emissions per unit of GNP than the US, and only Switzerland and Sweden have lower

Table 1.7. **CO_2 Emissions from Fossil Fuels in OECD Countries, 1987**

	Emissions (million Metric Tons)	Emissions Per Capita	Emissions Per $10 000 GNP
Australia	64	3.8	3.6
Austria	14	1.8	1.5
Belgium[1]	26	2.6	2.3
Canada	108	4.0	2.8
Denmark[1]	17	3.4	2.2
Finland	14	2.8	2.0
France[1]	91	1.6	1.3
Germany[1]	266	3.5	2.5
Greece[1]	14	1.4	3.5
Iceland	–	–	–
Ireland[1]	8	2.0	3.6
Italy[1]	97	1.7	1.6
Japan	238	1.9	1.2
Luxembourg[1]	2	5.6	2.9
Netherlands[1]	36	2.4	2.1
New Zealand	6	2.0	2.3
Norway	12	3.0	1.7
Portugal[1]	8	0.8	2.8
Spain[1]	44	1.1	1.9
Sweden	15	1.9	1.1
Switzerland	10	1.4	0.7
Turkey	34	0.6	5.3
UK[1]	155	2.7	2.6
US	1 212	4.9	2.7
EC	764	2.2	2.1
OECD	2 491	2.9	2.2

1. EC countries.
Source: Calculated from World Resources Institute (1990), Tables 24.1, 15.1 and 16.1.

emissions per unit of GNP than Japan. According to Nitze (1990), the Nordic countries, Australia and New Zealand would tend to go along with targets agreed by the EC. Canada has indicated its sympathy with objectives proposed by some EC countries, but its position more closely resembles that of the United States. Nitze contends that Japan would probably go along with whatever target the United States and the European Community agree on, but this does not seem to be a foregone conclusion; Japan's CO_2 emissions per unit of GNP are much lower.

Treating the EC countries as a unit would seem legitimate, since the EC can act as a signatory to a Convention (the EC is a signatory to the Montreal Protocol), and often negotiates on behalf of its member countries (as in the GATT negotiations; see Grieco, 1990). However, there are great differences between countries within the EC, as Table 1.7 shows. What is more, the stated policy objectives of the EC countries also vary. The Netherlands and Germany have made unilateral commitments to reduce emissions, whereas Spain, Portugal and Greece would prefer weaker targets because of their need for economic growth and their smaller historic contribution to the build up of atmospheric concentrations. Despite these differences, it seems conceivable, if not likely, that the EC countries will be able to act as a unit in negotiations (Nitze, 1990, p. 7):

"The European Community as a bloc will probably commit itself to internationally-agreed targets on CO_2 emissions limitation, with different member states accepting different sub-targets on the basis of internal negotiations similar to those which preceded issuance of the large-scale combustion plant directive."

It should be noted here that the Montreal Protocol accords the EC countries special status; they are treated as a single country for purposes of meeting the consumption reduction obligations of the Convention. This means that the Community is allowed some flexibility in deciding how the overall EC obligation would be best met by the individual member states. Hence, the Montreal Protocol would seem to support the assumption that the EC can be treated as a single negotiating unit.

The data underlying the analysis of OECD negotiations that follows are presented in Table 1.8.

Table 1.8. **Data for Hypothetical OECD Negotiations**

	US	EC	Japan	OECD
CO_2 Emissions [1]	1 212	764	238	2 491
CO_2 per capita	4.9	2.2	1.9	2.9
CO_2 per $10 000 GNP	2.7	2.1	1.2	2.2
GNP [2]	4 517	3 661	1 925	11 272

1. Million metric tons.
2. Billion US dollars.
Source: Compiled from World Resources Institute (1990), Tables 24.1, 15.1 and 16.1.

The simplest case

If the US, EC and Japan are identical in the sense that they each have the same b_i and c_i parameters, then the problem is simple to evaluate. The reason is that the negotiating positions of the three countries depend solely on the value of c/b, and not on the individual values taken by c_i and b_i (actual net benefits will depend on the *absolute* values of c and b, but only *relative* magnitudes matter in this analysis).

Table 1.9 summarizes the results. The Table indicates whether an agreement could be reached using the various instruments. Because all countries have identical cost parameters, and because marginal abatement costs are assumed to depend on per cent abatement, the marginal abatement costs of all three countries are identical when a uniform per cent abatement level is negotiated. Hence, this outcome is identical (both in terms of its "negotiating acceptability" and its cost-effectiveness) to the case where trading is allowed, and to the case where a uniform tax is imposed.

The analysis assumes that an agreement can be reached if all three parties can be made better off by negotiating with the given instrument than by reverting to the non co-operative outcome. The results indicate, perhaps surprisingly, that negotiation on the basis of uniform per cent abatement (equivalently, uniform per cent abatement with trading or a uniform carbon tax) is *always* acceptable to the three parties. One reason for this is the assumption that marginal abatement costs depend on per cent abatement. Given this assumption, uniform per cent abatement implies that each country will incur the same marginal abatement cost. Hence, trading is not necessary, and no transfers need take place. Another reason stems from the levels of both GNP and pre-control emissions. The US has high values for both, and hence bears a bigger share of the burden in the non co-operative outcome. In negotiating a uniform per cent abatement, the US increases its abatement and, in so doing, incurs a substantial cost. However, the US is compensated for this by much greater abatement by the EC and Japan. These last countries do not benefit as much from the greater abatement as the US, but they also do not have to bear as large a cost.

When c/b is "small" (9 or less in the Table), an allocation of emission permits on a per capita basis is acceptable to all three parties. The US is at a disadvantage when this instrument is used, because its emissions per capita are more than twice as large as those of the EC and Japan. However, the US still prefers this instrument to no agreement

Table 1.9. **Acceptability of Negotiation Instruments When b_i and c_i Parameters are Identical**

c/b	Uniform Per Cent Abatement	Uniform Per Capita Allocation	Uniform Per GNP Allocation
9 or less	x	x	x
10-13	x	–	x
14 or greater	x	–	–

1. Indicates that the instrument is acceptable to all three parties.
2. Indicates that the instrument is not acceptable to at least one party.
Source: Computed using the data in Table 1.8 and the algorithm described in the text.

34

because, while the US would buy permits from the EC and Japan, the cost of these permits is small in relation to the benefits received by the US. This is no longer the case when c/b is "large" (10 or greater).

Allocations based on GNP are acceptable to all three parties at higher values for c/b (up to 13 in the table), because differences in CO_2 per unit of GNP are smaller than are differences in CO_2 per capita. The US is still at a disadvantage, but the disadvantage is not as great as when emission levels are negotiated on a per capita basis.

In general, the US would prefer to negotiate on the basis of uniform per cent abatement (equivalently, uniform per cent abatement with trading or a uniform carbon tax); the EC on the basis of uniform CO_2 per capita; and Japan on the basis of uniform CO_2 per unit of GNP. However, the instrument that is preferred by one party may not be acceptable to one or both of the other negotiating parties. Whether countries will disagree strongly on the appropriate instrument depends on whether a country's claim to not accepting a particular instrument is credible. This will depend in turn on the value of c/b. The economic analysis of global warming by Nordhaus (1990) suggests that c/b is "large," in contrast to other environmental problems like ozone depletion (see Barrett, 1991a). If countries do in fact perceive c/b to be "large," then this analysis indicates that negotiations should focus on uniform per cent abatement obligations (with or without trading), or a uniform carbon tax.

To obtain a deeper understanding of the mechanics of this analysis, it will prove helpful to examine the details of a particular simulation. Table 1.10 presents the results of a simulation which assumes that $c = 35$ and $b = 1$[14]. The simulation illustrates a number of points.

In the full co-operative outcome, all three parties undertake the same percentage abatement. This is because marginal abatement costs are assumed to depend on percentage abatement, and the full co-operative outcome requires that marginal abatement costs be equal for all three countries. However, net benefits vary among countries because of differences in both base-year emission levels and GNP. If base-year emissions are higher for Country A than for Country B, then A will have higher total abatement costs, simply because it is abating more emissions, even if both A and B have the same marginal cost of abatement. Likewise, all countries benefit from the same level of global abatement, but countries with larger economies will receive large total benefits, because it is assumed that all countries have the same b_i parameter. Very roughly, the US receives lower net benefits than the EC, because the US has greater emissions per unit of GNP and greater emissions in total. The US abates more, and hence incurs a higher total cost. However, it also has a larger GNP, and hence receives a greater total benefit. But because its emissions per unit of GNP are much greater than the EC's, the US's higher cost (relative to the EC) exceeds its higher benefits (relative to the EC).

In the non co-operative outcome, the US bears a larger share of the burden of abatement in percentage terms. This is because the non co-operative outcome requires that each country set *its* marginal abatement benefit equal to *its* own marginal abatement cost. The US has a higher marginal abatement benefit for any given level of global abatement by virtue of its higher GNP. Hence, the US abates a greater percentage of its emissions, because it benefits more from abatement at the margin than the other countries, while all countries have the same marginal cost of abatement.

One can calculate the uniform level at which a country would *prefer* that a given instrument be employed by allowing that country to choose a level which maximises its

Table 1.10. **Simulation Results for Case Where** $c = 35, b = 1$

	US	EC	Japan	Total
Full Co-operative Outcome				
Per Cent Abatement	25.9%	25.9%	25.9%	25.9%
Net Benefits	1 034	1 094	767	2 895
Non Co-operative Outcome				
Per Cent Abatement	12.4%	10.0%	5.3%	10.8%
Net Benefits	731	721	438	1 891
Uniform Per Cent Abatement				
Preferred Choice	21.6%	27.0%	42.5%	–
Net Benefits @ 21.6%	1 078	1 051	686	2 814
Net Benefits @ 27.0%	1 007	1 096	786	2 889
Net Benefits @ 42.5%	60	738	905	1 703
Uniform Per Capita Allocation				
Preferred Choice[1]	–	2.0	1.7	–
Per Cent Abatement	–	36.6%	46.3%	–
Net Benefits @ 2.0	–3 110	3 428	2 083	2 401
Net Benefits @ 1.7	–4 271	3 186	2 180	1 095
Uniform Per GNP Allocation				
Preferred Choice[2]	2.0	1.6	1.2	–
Per Cent Abatement	7.2%	28.9%	46.7%	–
Net Benefits @ 2.0	80	596	712	1 389
Net Benefits @ 1.6	–647	1 366	2 136	2 855
Net Benefits @ 1.2	–2 327	850	2 498	1 021

1. Allowed emissions per capita.
2. Allowed emissions per $10 000 GNP.

net benefits, on the assumption that the other countries undertake the chosen level. For example, Table 1.10 indicates that if negotiations focused on uniform per GNP allocations, then the US would prefer that this level be 2 metric tons of CO_2 per $10 000 GNP, whereas the EC would prefer a level of 1.6 and Japan of 1.2. Given these calculations, one can determine which *instrument* a country would prefer by observing which instrument (at the country's preferred level) delivers that country the greatest net benefit.

Table 1.10 indicates that each country prefers to negotiate on the basis of a different instrument. The US would like to negotiate on the basis of uniform percentage abatement, for it is only using this instrument that the US can earn higher net benefits compared to the non co-operative outcome. The EC would prefer to negotiate on a uniform per capita allocation of permits, but will accept negotiations on the other two instruments. Japan would prefer to negotiate on a uniform per GNP allocation, but, like the EC, would be willing to negotiate on the other instruments. Only uniform percentage abatement is acceptable to all three parties.

The US would prefer a uniform per cent abatement of 21.6 per cent. This level of abatement makes all three parties better off compared with the non co-operative outcome, and hence may prove acceptable. The EC's preferred abatement of 27.0 per cent also makes all three parties better off compared with the non co-operative outcome. However, the US would prefer a smaller percentage abatement, while Japan would prefer a larger percentage abatement. Japan's preferred choice for a uniform percentage abatement level would not be acceptable to the US, because the US would do better without an agreement. However, Japan is likely to argue that the EC's preferred choice is better than that of the US. Notice that a negotiated percentage abatement close to the EC's first choice would make the whole group of countries nearly as well off as in the full co-operative case.

No allocation of emission permits on a per capita basis would be acceptable to the US. The levels preferred by the EC and Japan would require "too much" abatement. Japan's preferred choice for a per capita allocation would actually result in lower total net benefits compared with the non co-operative outcome.

Negotiations on the basis of emissions per unit of GNP are also not acceptable to the US, since the US can always do better by pulling out of the negotiations. Again, Japan's preferred choice for a level of emissions per unit of GNP actually results in lower total net benefits compared with the non co-operative case.

More complicated cases

The preceding analysis assumes that all OECD countries are identical with respect to their b_i and c_i parameters. It is likely that these parameters will, however, vary from one country to another for reasons explained earlier. Not much is known about these parameter values – or, more generally, about the nature of costs and benefits for individual countries. Rather than make specific assumptions about these values, the following analysis considers a number of cases.

Each of the cases takes $b_i = 1$ for all countries, but varies the values for c_i. Table 1.11 lists the cases considered. In these cases, c_i takes on a value of 35 or a value 50 per cent larger (52.5). Otherwise, the analysis is identical to the one carried out previously.

The results are summarized in Table 1.12. Despite the differences in costs, the per capita and per GNP allocation rules are still unacceptable to all three parties. The reason seems to be that c_i/b_i remain high in this analysis.

In all the cases examined, all three countries prefer the rule of a uniform percentage abatement with trading, to the rule where percentage abatement is uniform, but trading is prohibited.

In most cases, the high-cost countries prefer the carbon tax, and the low-cost countries prefer uniform percentage abatement with trading. The reason is that the tax does not require as much abatement on the part of high-cost countries, while uniform abatement with trading allows the low-cost countries to sell permits to the high-cost countries. However, there are two important exceptions to this. In case 7, Japan prefers to trade, even though it is a high-cost country. In case 2, Japan prefers the tax, even though it is a low-cost country. It seems that Japan's preferred instrument in these cases is influenced by what the US does. In case 7, the tax that is acceptable to the US is quite small, so that while this tax keeps Japan's costs down, it also keeps its benefits down. In case 2, the trading scheme that is acceptable to the US does not reduce emissions by very much, and so benefits are reduced. The US will accept a carbon tax that achieves more

Table 1.11. **Simulation Cases: Values for the Parameters c_i**

Case	US	EC	Japan
1	35	35	35
2	52.5	35	35
3	35	52.5	35
4	35	35	52.5
5	52.5	52.5	35
6	52.5	35	52.5
7	35	52.5	52.5
8	52.5	52.5	52.5

abatement, and that is why Japan prefers the tax in this case. This finding underscores the importance of taking behavioural reactions into account.

This analysis considers only the position of individual countries – and, in particular, whether a country's net benefits will be greater using one instrument rather than another, given the (best) responses by the other negotiating parties. It does not consider the *sum* of net benefits for the group. Yet, it would be interesting to know whether, among the instruments that are acceptable to negotiators, one improves *total* net benefits more than the others. For the six cases in Table 1.12 where the c_i vary, it turns out that the tax instrument resulted in higher total net benefits in three cases, and the acceptable trading scheme resulted in higher total net benefits in the other three cases. Hence, it seems that one would need to know a lot about the nature of net benefits for all negotiating parties to be able to say whether one instrument would achieve a better result in total than the other.

Table 1.12. **Summary of Simulation Analysis for Cases Identified in Table 1.11 (with $b_i = 1$)**

Instrument	1	2	3	4	5	6	7	8
Uniform Per Cent Abatement No Trading	x*	x	x	x	x	x	x	x*
Uniform Per Cent Abatement With Trading	x*	x^2	x^{1,3}	x^{1,2}	x^3	x^2	x^{1,3}	x*
Uniform Carbon Tax	x*	x^{1,3}	x^2	x^3	x^{1,2}	x^{1,3}	x^2	x*
Uniform Per Capita Allocation	–	–	–	–	–	–	–	–
Uniform Per GNP Allocation	–	–	–	–	–	–	–	–

x   Indicates instrument acceptable to all three parties.
–   Indicates instrument not acceptable to at least one party.
*   Indicates that all instruments are equally preferred by all three parties.
1. Indicates instrument preferred by US.
2. Indicates instrument preferred by EC.
3. Indicates instrument preferred by Japan.

Additional simulations which vary the absolute magnitude of the c_i parameters, but retain the relative ranking in Table 1.11 (while still holding $b_i = 1$ for all i) indicate that many of the results for the simple case where every country had the same c_i and b_i parameters still hold. As long as the c_i parameters are "small," differences in their relative magnitude don't matter very much for negotiations. All the negotiating instruments are acceptable to all the parties. When all the c_i are "large," only the uniform per cent abatement with trading and carbon tax schemes are acceptable to all three parties. As noted earlier, current information suggests that the c_i parameters are likely to be large in the case of global warming – at least for fairly substantial levels of abatement. Hence, these instruments would appear most attractive for OECD negotiations.

To provide a greater understanding of the nature of these results, Table 1.13 summarizes the findings of one simulation (case 6 in Table 1.11).

In the full co-operative outcome, the EC undertakes greater percentage abatement, because its marginal abatement costs are lower than the other two countries for any given percentage abatement.

In the non co-operative outcome, the US no longer abates a greater percentage of emissions than the EC. This is because the effect of higher costs offsets the effect of receiving greater benefits from any given level of total abatement.

Of the three choices for a uniform per cent abatement (without trading), only the US's first choice is acceptable to all three parties; the others would make the US worse off compared to the non co-operative outcome, and so would not be acceptable to the US.

When trading is allowed, all countries would prefer a greater percentage abatement. The reason is that trading lowers the costs of complying with an obligation, and hence allows countries to accept a slightly greater burden. Notice that in all of the three trading cases, the US and Japan buy permits from the EC. This is simply because the EC has lower marginal abatement costs at any given uniform percentage reduction.

Despite the fact that the US and Japan have identical marginal abatement cost functions, Japan prefers a much greater carbon tax. The reason is that the higher tax will result in much more abatement (in absolute terms) by the US and the EC, and the additional benefits outweigh the increase in Japan's abatement costs.

There is no uniform per capita allowance that would be acceptable the US. The EC and Japan would benefit greatly from such a scheme, were it acceptable to the US. However, a refusal by the US to negotiate using this instrument would be credible. The same is true of a "per GNP" allocation.

Of the remaining instruments, notice that a 15.1 per cent uniform abatement with trading is superior to a 14.8 per cent uniform abatement without trading; the former is strictly preferred by all parties to the latter. All three countries would negotiate on a uniform abatement with trading scheme (at around 15 per cent) or on a carbon tax (of about 8 – the units haven't been specified in this exercise). However, the EC would prefer the trading scheme, because it could make around 900 from this scheme but less than 700 with the tax. The US and Japan would prefer the tax. The US would get about 750 under the trading scheme, but close to 1 000 under the tax. Japan would receive about 485 with trading, but close to 600 with the carbon tax.

Finally, notice that in this example, the trading scheme would result in greater total net benefits. If the 15 per cent uniform abatement with trading scheme were accepted,

Table 1.13. **Simulation Results for Case 6 ($b_i = 1$)**

	US	EC	Japan	Total
Full Co-operative Outcome				
Per Cent Abatement	17.7%	26.5%	17.7%	20.7%
Net Benefits	993	671	651	2 315
Non Co-operative Outcome				
Per Cent Abatement	8.3%	10.1%	3.5%	8.4%
Net Benefits	608	535	345	1 488
Uniform Per Cent Abatement (No Trading)				
Preferred Choice	14.8%	27.0%	30.0%	–
Net Benefits @ 14.8%	740	871	475	2 086
Net Benefits @ 27.0%	232	1 096	634	1 961
Net Benefits @ 30.0%	–45	1 083	640	1 677
Uniform Per Cent Abatement (With Trading)				
Preferred Choice	15.1%	29.1%	30.6%	–
Actual Abatement @15.1%	12.9%	19.3%	12.9%	15.1%
Net Benefits @ 15.1%	755	906	485	2 146
Actual Abatement @29.1%	24.8%	37.2%	24.8%	29.1%
Net Benefits @ 29.1%	109	1 178	651	1 937
Actual Abatement @30.6%	26.1%	39.2%	26.1%	30.6%
Net Benefits @ 30.6%	–42	1 175	652	1 785
Uniform Carbon Tax				
Preferred Choice	8.9	7.7	17.7	–
Actual Abatement @ 8.9	17.0%	25.4%	17.0%	19.9%
Net Benefits @ 8.9	995	683	634	2 312
Actual Abatement @ 7.7	14.7%	22.1%	14.7%	17.2%
Net Benefits @ 7.7	977	699	575	2 251
Actual Abatement @ 17.7	33.7%	50.5%	33.7%	39.5%
Net Benefits @ 17.7	28	–465	842	405
Uniform Per Capita Allocation				
Preferred Choice	–	1.9	1.8	–
Abatement @ 1.9 (No Trading)	62.0%	15.3%	1.9%	39.4%
Abatement @ 1.9 (With Trading)	33.6%	50.4%	33.6%	39.4%
Net Benefits @ 1.9 (With Trading)	–6 027	4 279	2 173	425
Abatement @ 1.8 (No Trading)	62.4%	16.2%	2.9%	40.0%
Abatement @ 1.8 (With Trading)	34.1%	51.2%	34.1%	40.0%
Net Benefits @ 1.8 (With Trading)	–6 157	4 278	2 173	295
Uniform Per GNP Allocation				
Preferred Choice	2.2	1.5	1.3	–
Abatement @ 2.2 (No Trading)	18.4%	–4.9%	–77.0%	0.1%
Abatement @ 2.2 (With Trading)	0.1%	0.2%	0.1%	0.1%
Net Benefits @ 2.2 (With Trading)	–	12	15	26
Abatement @ 1.5 (No Trading)	43.9%	27.9%	–21.7%	31.4%
Abatement @ 1.5 (With Trading)	26.7%	40.1%	26.7%	31.4%
Net Benefits @ 1.5 (With Trading)	–2 261	1 540	2 424	1 702
Abatement @ 1.3 (No Trading)	51.9%	38.2%	4.4%	41.1%
Abatement @ 1.3 (With Trading)	35.1%	52.6%	35.1%	41.1%
Net Benefits @ 1.3 (With Trading)	–3 895	1 390	2 568	64

total net benefits would be around 2 150. If a tax of between 8 and 9 units were accepted, total net benefits would be between 2 250 and 2 300.

Hypothetical non-OECD negotiations

Table 1.14 provides data on the three most important non-OECD countries from an emissions perspective. As noted earlier, agreement by these countries to reduce emissions would be needed if global abatement were to be significant, especially in the longer term.

Simulation analysis using a wide range of values for b_i and c_i (assumed constant for each country; see the discussion below) show that agreement within this group of countries and between this group and the OECD cannot be reached using *any* instrument.

Even agreement between the US, the other OECD countries and the former USSR cannot be reached in the simulations. One reason for this is that the former USSR has a high level of emissions, given its GNP. If abatement is high, the former USSR benefits because of its relatively large GNP. However, because of its large emission level, the former USSR also has to incur a substantial total cost of abatement. The OECD countries are not in this position. Another reason is that in the non co-operative outcome, the former USSR benefits from a much greater percentage abatement taken by the other countries, while the former USSR undertakes little abatement itself. Any agreement using the instruments considered in this paper would demand much more abatement on the part of the former USSR, even if the OECD countries also undertake more abatement. Hence, free-riding is likely to be attractive to the former USSR.

Agreement between the former USSR, China and Brazil also proves elusive. In these cases, the incentive which the former USSR faces to free-ride is very small, because China and Brazil do not abate emissions by much in the non co-operative outcome. The ex-USSR would like to negotiate an agreement, because it could benefit by greater abatement by the other countries. However, China and Brazil are only made worse off using any of the available instruments. The reason is that CO_2 per unit of GNP is extremely large for these countries. The benefit of an agreement to reduce emissions is overwhelmed by the huge cost which these countries would have to bear in abating their emissions.

Table 1.15 presents the results of one simulation involving the OECD, the former USSR, and China. The Table indicates that the OECD would agree to uniform per cent abatement, but not the ex-USSR and China; they do better with no agreement at all.

Table 1.14. **Data for Hypothetical Non-OECD Negotiations, 1987**

	ex-USSR	China	Brazil[1]
CO_2 Emissions	1 015	572	1 250
CO_2 Per Capita	3.5	0.5	8.3
CO_2 Per $10 000 GNP	4.3	18.5	43.1
GNP	2 357	314	286

1. Includes emissions due to deforestation.
Source: Calculated from World Resources Institute (1990), Tables 24.1, 15.1 and 16.1.

Table 1.15. **Simulation Results for Case Where** $c = 35$, $b = 1$
OECD, USSR, and China

	OECD	ex-USSR	China	Total
Full Co-operative Outcome				
Per Cent Abatement	30.8%	30.8%	30.8%	30.8%
Net Benefits	8 422	940	–601	8 761
Non Co-operative Outcome				
Per Cent Abatement	27.8%	5.8%	0.8%	18.5%
Net Benefits	4 572	1 601	221	6 394
Uniform Per Cent Abatement				
Preferred Choice	38.0%	22.6%	6.1%	–
Net Benefits @ 38.0%	8 734	578	–1 027	8 285
Net Benefits @ 22.6%	7 294	1 085	–245	8 134
Net Benefits @ 6.1%	2 582	507	39	3 129
Uniform Per Capita Allocation				
Preferred Choice	1.4	–	1.0	–
Per Cent Abatement	22.9%	–	42.5%	–
Net Benefits @ 1.4	1 402	–1 999	8 778	8 181
Net Benefits @ 1.0	368	–4 036	11 161	7 493
Uniform Per GNP Allocation				
Preferred Choice	1.7	–	–	–
Per Cent Abatement	41.8%	–	–	–
Net Benefits @ 1.7	15 508	–2 477	–5 386	7 645

China would like to negotiate on the basis of a uniform per capita allocation of permits, but this instrument can only leave the OECD and ex-USSR worse off. A uniform per GNP allocation is only acceptable to the OECD countries.

Table 1.16 presents the results of another simulation involving negotiations between the OECD, China and Brazil (where Brazil's emissions include those arising from deforestation). Again, agreement cannot be reached. The OECD would negotiate on the basis of uniform per cent abatement, but this instrument can only make China and Brazil worse off. A uniform per capita allocation is acceptable to China, but not to the OECD and Brazil. A uniform emissions per GNP allocation is acceptable only to the OECD.

These simulations all assume that all countries have the same b_i and c_i parameters. However, as explained in the previous section, one would expect that c_i/b_i would, if anything, be higher for the non-OECD countries. Hence, any alteration in the relative parameter values will serve to make agreement even harder to reach.

The implications of this analysis are clear. Many non-OECD countries will not participate in a Convention unless they are allowed to bear a smaller relative burden. This is partly because the characteristics of these countries – including high emissions per unit of GNP – provide them with a weak incentive to abate emissions in the event that co-operation fails. The OECD countries have a greater incentive to abate emissions if

Table 1.16. **Simulation Results for Case Where** $c = 35$, $b = 1$
OECD, China, and Brazil

	OECD	China	Brazil	Total
Full Co-operative Outcome				
Per Cent Abatement	26.8%	26.8%	26.8%	26.8%
Net Benefits	8 548	−396	−1 280	6 873
Non Co-operative Outcome				
Per Cent Abatement	28.1%	0.8%	0.7%	16.5%
Net Benefits	4 082	209	190	4 480
Uniform Per Cent Abatement				
Preferred Choice	38.9%	6.4%	2.8%	–
Net Benefits @ 38.9%	9 457	−1 068	−2 903	5 486
Net Benefits @ 6.4%	2 866	44	−13	2 897
Net Benefits @ 2.8%	1 294	29	17	1 340
Uniform Per Capita Allocation				
Preferred Choice	1.4	1.1	–	–
Per Cent Abatement	30.5%	43.5%	–	–
Net Benefits @ 1.4	3 297	12 182	−8 733	6 747
Net Benefits @ 1.1	2 694	13 383	11 853	4 225
Uniform Per GNP Allocation				
Preferred Choice	2.0	–	–	–
Per Cent Abatement	44.2%	–	–	–
Net Benefits @ 2.0	23 128	−5 408	−13 705	4 015

negotiations fail. Hence, if agreement cannot be reached, the non-OECD countries benefit from abatement by the OECD countries without having to undertake much expensive abatement themselves; the non-OECD countries, therefore, have a strong incentive to free-ride.

Another problem is that the instruments considered here serve to polarise the two groups of countries. Uniform per cent abatement is attractive to the OECD countries, but not to the non-OECD countries. A uniform per capita allocation appeals to some – again, not all – non-OECD countries, but appears not to be acceptable to the OECD. Negotiations between the OECD and non-OECD would prove more successful if an alternative instrument could be found which did not so obviously favour one group over another. However, a simple instrument that achieves this objective may not exist. What is more, if such an instrument could be found, it may serve to undermine the overall achievement of negotiations – for example, it may win agreement between the OECD and non-OECD, but not prove as successful in intra-OECD negotiations as the blunter instruments.

An alternative, and perhaps more fruitful, approach would be for the non-OECD countries to be allowed to meet weaker obligations, and for the OECD countries to assist these countries by making side payments that compensated the non-OECD countries for the incremental cost of abatement. Under the tradeable permit schemes considered here, it has been assumed that a competitive market in permits develops, and that permits trade at the market-clearing price. Since the marginal abatement cost schedules are upward-sloping, this means that the countries that sell permits earn a surplus over and above the cost of undertaking abatement on behalf of the countries that buy permits. Of course, the countries that buy permits also earn a surplus, because the cost to them of abating the same emissions would be greater than the market price of permits (their marginal abatement cost curves are also upward-sloping). However, the simulations show that this surplus is not sufficient to make the OECD countries want to participate in negotiations. Negotiations would prove more successful if the surplus being received by the non-OECD countries could be reduced. The point is that negotiations can only succeed if all parties are made better off with agreement than without it. Paradoxically, reducing the transfers to non-OECD countries can actually make these countries better off.

These last issues – deterring free-riding and designing acceptable side payment mechanisms – are the subject of the next two Chapters.

Notes

1. Manne and Richels (1990) estimate that OECD carbon emissions are 55 per cent of the total in 1990, but only 31 per cent in 2100, if no action is taken to abate emissions.

2. For an estimate of the loss associated with using only a gasoline tax, see Nordhaus (1990).

3. Barrett (1991*d*) investigates the negotiation of an European Community carbon tax using a methodology similar to the one employed here. However, Barrett (1991*d*) employs non-linear abatement cost functions.

4. This is a common assumption in global analyses of climate change policy. See, for example, International Energy Agency (1989), Manne and Richels (1990), and Whalley and Wigle (1990).

5. Barrett (1991*d*) carries out a similar analysis, but using a horizontal marginal damage curve.

6. Current evidence suggests that agriculture is not vulnerable in total; some areas will lose, others gain, but the total effects are expected to be small. See Nordhaus (1990) and Crosson (1989).

7. GNP is an imperfect measure of potential damage from climate change for a number of reasons. Chief among these is the fact that GNP does not reflect the value of non-marketed goods and services, which may be profoundly affected by climate change (see Nordhaus, 1990).

8. A country may care about another's per capita consumption, but try to improve this through development assistance, rather than through a climate change Framework Convention.

9. However, the US has undertaken a number of policies, which it claims will reduce net greenhouse gas emissions in the year 2000 to a level no greater than the 1987 level. See Reinstein (1991).

10. Previous negotiations on environmental issues have had difficulty settling on a base year, but not on an end year (an example being the sulphur and nitrogen oxides protocols to the Long Range Transboundary Convention).

11. The schedule of annual emission levels could be calculated by minimising the present value sum of abatement plus adjustment costs subject to the constraint that the target be met in the specified year. If there were no adjustment costs, then the optimal path would involve no abatement until the specified year, and then only that amount of abatement which was necessary to just meet the target. Where there are adjustment costs, the optimal path would be gradual. Common sense suggests that the path should be gradual.

12. Trading is not allowed under the Sulphur and Nitrogen Oxides Protocols to the Long Range Transboundary Air Pollution Convention, but these agreements were intended to reduce acid rain emissions, and for this problem, the location of polluters matters. Unrestricted trading may have meant that emissions would have been abated least in the areas where abatement was needed most.

13. One reason for this may be that the pay-offs that countries would get if negotiations failed are themselves non-uniform. If countries are identical in every respect except for their base-year

emission level and the GNP, then it can be shown that a country with a higher product of these two values will bear a larger share of the burden (defined as pre-control abatement costs divided by GNP).

14. Recall that the slope of the marginal abatement benefit function is b_i*G_i. In the simulation analysis, GNP is taken to be 4 517 for the US, 3 661 for the EC and 1 925 for Japan. Hence, the slopes of the marginal abatement benefit functions are larger than 1 for all three countries.

References

BARRETT, S. (1991a). *The Paradox of International Environmental Agreements,* mimeo, London Business School.

―――, S. (1991b). "Economic Analysis of International Environmental Agreements: Lessons for a Global Warming Treaty," in OECD, *Responding to Climate Change: Selected Economic Issues,* Paris: OECD.

―――, S. (1991c). "Economic Instruments for Climate Change Policy," in OECD, *Responding to Climate Change: Selected Economic Issues,* Paris: OECD.

―――, S. (1991d). *Reaching a CO_2 Emission Limitation Agreement for the Community: Implications for Equity and Cost-Effectiveness.* Paper prepared for the European Commission, Brussels.

BURTRAW, D. and M.A. Toman (1991). *Equity and International Agreements for CO_2 Containment,* Resources for the Future Discussion Paper ENR91-07, Washington, DC.

GOLLOP, F.M. and M.J. Roberts (1983). "Environmental Regulations and Productivity Growth: The Case of Fossil-fueled Electric Power Generation," *Journal of Political Economy,* 91: 654-673.

GRIECO, J.M. (1990). *Cooperation Among Nations: Europe, America, and Non-Tariff Barriers to Trade,* Ithaca: Cornell University Press.

HAHN, R.W. (1989). "Economic Prescriptions for Environmental Problems: How the Patient Followed the Doctor's Orders," *Journal of Economic Perspectives,* 3: 95-114.

INTERNATIONAL ENERGY AGENCY (1989). *Policy Measures and their Impact on CO_2 Emissions and Accumulations,* Paris: International Energy Agency.

IPCC (1990). *Scientific Assessment of Climate Change,* Executive Summary, 25 May.

JORGENSON, D.W. and P.J. Wilcoxen (1989). *Environmental Regulation and U.S. Economic Growth,* Harvard Institute of Economic Research Discussion Paper No. 1458, October.

KEOHANE, R.O. (1984). *After Hegemony: Cooperation and Discord in the World Political Economy,* Princeton: Princeton University Press.

MANNE, A.S. and R.G. Richels (1990a). *Global CO_2 Emission Reductions – the Impacts of Rising Energy Costs,* mimeo, Stanford University, February.

―――, A.S. and R.G. Richels (1990b). *International Trade in Carbon Emission Rights: A Decomposition Procedure,* mimeo, Stanford University, November.

NITZE, W.A. (1990). *The Greenhouse Effect: Formulating a Convention,* London: Royal Institute of International Affairs.

NORDHAUS, W.D. (1990). *To Slow or Not to Slow: The Economics of the Greenhouse Effect,* mimeo, Department of Economics, Yale University, February 5.

PEARCE, D. (1990). *Greenhouse Gas Agreements: Part 1, Internationally Tradeable Greenhouse Gas Permits,* mimeo, Department of Economics, University College London.

———, D. (1991). "The Global Commons," in D. Pearce (ed.) *Blueprint 2: Greening the World Economy,* London: Earthscan.

RAIFFA, H. (1982). *The Art and Science of Negotiation,* Cambridge, Mass: Harvard University Press.

REINSTEIN, R. (1991). *Prepared Testimony, House of Representatives,* 21 February.

SCHELLING, T. (1983), "Climate Change: Implications for Welfare and Policy," in *Changing Climate: Report of the Carbon Dioxide Assessment Committee,* Washington, DC: National Academy of Sciences.

WHALLEY, J. and R. Wigle (1989). *Cutting CO_2 Emissions: The Effects of Alternative Policy Approaches,* mimeo, University of Western Ontario.

———, J. and R. Wigle (1990). *The International Incidence of Carbon Taxes.* Paper prepared for "Economic Policy Responses to Global Warming," Conference organized by the Instituto Bancario Sao Paolo di Torino, October 1990.

WORLD BANK (1990). *World Development Report 1990,* Oxford: Oxford University Press.

WORLD RESOURCES INSTITUTE (1990). *World Resources 1990-91,* Oxford: Oxford University Press.

Chapter 2

Side payments in a global warming convention

INTRODUCTION AND SUMMARY

"The questions of how much financial assistance should be given to developing countries in carrying out their obligations under a convention, through what mechanisms such assistance should be provided, and how it should be allocated will be the most challenging and potentially divisive issues in negotiating a greenhouse convention."

Nitze (1990, p. 48)

The idea that side payments would be needed to make a global warming Framework Convention truly effective has been accepted as a matter of principle by many, if not most, potential negotiating parties. The reason is basically that an effective Convention must include a great number of countries, and many of these will be made worse off in joining a Convention unless compensated by the parties that would be made better off. What has *not* been accepted, as Nitze observes, is how large the payments should be in total, and how much should particular countries contribute to, or receive from, this total.

Side payments have been employed to resolve many international environmental disputes. A few examples will help illustrate the importance of side payments in reaching agreement.

The North Pacific Fur Seal Treaty, originally negotiated in 1911 but still in force today, banned the wasteful practice of hunting seals at sea. In restricting harvesting to land, total product was increased because land harvests were far more efficient. What is more, in banning pelagic sealing, the inefficiencies of open access were also eliminated, for the two countries with jurisdiction over the breeding grounds, the United States and Russia, had strong incentives to manage their stocks as "sole owners". However, the agreement to ban pelagic sealing would have harmed the two pelagic sealing nations, Canada and Japan, and if these countries did not agree to the treaty, the value of the agreement to the United States and Russia would be nil. To win the agreement of the pelagic sealing nations, the United States and Russia agreed to give Canada and Japan a fixed share of their annual catch[1].

The World Heritage Convention, negotiated in 1972 and signed by more than 90 countries, is intended to protect natural and cultural areas of "outstanding universal value." In recognising that natural areas such as the Serengeti and the Galapagos Islands

49

form a part of the world's common heritage, but that the cost of protection fall on the countries having jurisdiction over these areas, the treaty established the World Heritage Fund to help pay for the costs of protection. The Fund is small – $741 000 in 1984 – and assistance is not generous – the treaty generally insists that "only part of the cost of work necessary shall be borne by the international community." However, in establishing the Fund, the treaty has made all countries somewhat better off by helping to preserve natural environments that might otherwise have been exploited[2].

Perhaps the most relevant precedent is the Multilateral Fund established in June 1990 under Article 10 of the amended Montreal Protocol. The original Protocol, agreed in 1987, won approval from most industrial countries and a substantial number of developing countries – the latter being attracted by weaker obligations and the possibility of financial assistance. However, the original Protocol did not secure the approval of certain key developing countries, especially China and India, whose consumption of CFCs was expected to climb rapidly. It was to win the signatures of these countries that the Protocol was amended to include the Multilateral Fund, which is intended to meet all *incremental costs* of compliance by developing countries[3]. While much uncertainty surrounds the total amount of transfers required, it is generally agreed that the total cost will not be very large. One study estimated that the present value cost (discounted at a 5 per cent real rate) over the period 1990 - 2008 would be about $1.8 billion (Markandya, 1991).

The magnitude of transfers required to reach the full co-operative outcome – the outcome where global net benefits are maximised – will depend on the costs and benefits of abating greenhouse gas emissions. Many proposals for emission ceilings will eventually prove very costly. Although estimates vary and are subject to much uncertainty, current research suggests that the costs could range from 2-10 per cent of GNP by 2050, depending on the emission reduction target (Darmstadter, 1991). As shown later in this chapter, these figures suggest that the side payments needed to achieve the proposed targets could eventually be in the range of tens, if not hundreds, of billions of dollars *annually* (see also Manne and Richels, 1990).

However, these targets were not based on an analysis of costs and benefits. Analysis by Nordhaus (1990) suggests that the efficient level of greenhouse gas emission reduction – the level that equates the marginal costs of abatement with the marginal benefits – would cost less than $6 billion in 2050. One reason for the difference is that Nordhaus finds that only a modest reduction in CO_2 emissions can be justified economically – 6 per cent in 2050. Analysis by Manne and Richels (1990) suggest that costs would be much greater – 1-5 per cent of GNP, depending on the region, by 2030 – but they also assume that the level of abatement will be much greater – close to 50 per cent in 2030.

Perhaps surprisingly, the magnitude of the side payments required has no effect on the choice of a *mechanism* for making the side payments – at least not in theory. The real problem posed by the high costs of abatement, as explained in the next chapter, is that they make it difficult, if not impossible, to effectively deter free-riding. Hence, while the global warming problem would seem to be very different from that of conserving fur seals, protecting natural heritage, or saving the ozone layer, the mechanisms used in these treaties could offer lessons for designing a mechanism to collect and disperse side payments in a global warming Convention.

The North Pacific Fur Seal Treaty essentially establishes a rule for distributing the total net gains from full co-operation. Agreement by all parties requires that each receive an amount which is at least as great as that party could get by withdrawing from the treaty (either on its own or as part of a bigger coalition). For any particular problem, there will exist more than one allocation that would win unanimous approval. One possible rule would be to reward each country according to its contribution to the total net gain. While this mechanism has many attractive features, implementation requires knowledge of each party's net benefits. Such a calculation would be relatively easy in the case of sealing, for one can calculate the profits from harvesting (see Paterson and Wilen, 1977), and the contributions of each party to the grand coalition, since the number of countries involved is small. As demonstrated in the next section, such a rule could also be employed in the case of global warming – in theory. However, the net benefits associated with abating greenhouse gases will not be easy to calculate. Countries will have difficulty working out their own net benefits, never mind those of other countries. What is more, the number of parties involved will be large. Hence, negotiations will probably search for a simpler mechanism, one that would be easy to calculate, but which would also have broad appeal. Indeed, it may be for these reasons that the World Heritage Convention and Montreal Protocol decided on a different side payment mechanism.

The approach taken in these two agreements varies, however. The World Heritage Convention fixes the contribution to the Fund by each party (each party contributes one percent of its contribution to UNESCO every other year), and then allocates this amount to recipients on the basis of priorities established by a set of agreed "operational guidelines." The Montreal Protocol establishes criteria for compensation (incremental cost), and then allocates the burden for paying the compensation to donor countries on the basis of an agreed rule (either the United Nations scale of assessments or the 1986 level of CFC consumption). Both of these features could prove attractive in a global warming Framework Convention.

This chapter is organised into three sections. The first discusses the theory of bargaining with side payments. The second analyzes some proposals for side payment mechanisms in a global warming Convention. The final section raises the idea of linking issues to win broad agreement, rather than making monetary side payments. The main conclusions of the chapter are as follows:

1. In theory, an arbitrator could divide the total net gains from full co-operation so as to achieve agreement by all parties. However, calculation of such a division will be highly contentious in practice because countries will disagree on what their net benefits actually are. It is more likely that negotiators will search for simple mechanisms that also have the effect of rewarding most the countries that contribute most to the net benefits of the "grand coalition."

2. The net benefits to joining a Framework Convention will depend on all elements of the Convention and not just on the side payment mechanism. They will also depend on how attractive the Convention is to other potential signatories, and hence on what the Convention can be expected to accomplish.

3. Proposals to allocate tradeable emission permits on the basis of population will almost certainly fail to win agreement because the donors (countries with high emissions per capita) would be worse off with the agreement than without it. The great difficulty with this proposal is that the allocation rule provides recipient countries with an excessive reward for signing the Framework Convention. The reward is excessive because donors are required to give more than they benefit from having the recipients as signatories to the Convention, even while recipients receive more than they would require in order to be better off in the agreement

than outside it. The trading proposal would appeal more to the OECD countries if permits were allocated initially on the basis of historical emission levels. To make accession by non-OECD countries attractive, the OECD countries could offer to pay "incremental" costs.

4. Proposals to impose an international carbon tax are similarly flawed. The basic problem is that the revenue collected from such a tax will not bear any relation to the magnitude of the cost borne by non-OECD countries. An alternative proposal would be for all signatories to agree to impose the same tax nationally, but for the OECD countries to pay the "incremental costs" of non-OECD country accession.

5. Money side payments can be reduced by linking issues in negotiation which are mutually beneficial to the OECD and non-OECD countries.

THEORY

Global warming is similar to many social and economic problems in that the outcome that is best for all countries collectively clashes with the private incentives of individual countries. The most preferred outcome for country A is one where other countries abate their emissions but A does not. The same is true for countries B, C, D, etc. All countries want to be free riders. But if all countries do not abate their emissions, each is made worse off.

Global net benefits are maximised when countries co-operate fully, and this best collective outcome can be attained if:

i) commitments can be made binding;

ii) the negotiation instrument allows global abatement to be achieved at minimum cost; and

iii) side payments are feasible.

If commitments can be made binding, then free rider incentives can be eliminated. However, international law prevents commitments between countries from being enforced by an independent body (Barrett, 1990). Commitments between countries cannot be made legally binding; international agreements must be self-enforcing. Hence, mechanisms must be found which deter free-riding and yet are also self-enforcing. This is the subject of Chapter 3. The analysis of side payments presented here, like the analysis presented in Chapter 1, assumes that binding commitments *can* be made.

Cost-effective global abatement requires that the marginal abatement costs of all countries be equal. Chapter 1 already demonstrated that cost-effective abatement will be achieved if Convention obligations involve emission ceilings that can be traded internationally or an international emission tax, where revenues from the latter instrument were assumed to be retained by the national governments. However, Chapter 1 also demonstrated that all countries will not gain from such an agreement. Non-uniform obligations can win wider acceptance. But the full co-operative outcome could not be achieved without side payments.

The need for side payments

To see this, consider Table 2.1, which was derived from the simulation underlying Table 1.15. The full co-operative outcome demands that total emissions for the OECD, the former USSR and China fall 30.8 per cent. If this obligation is imposed uniformly, the OECD comes out much better than in the non co-operative outcome, but the ex-USSR and China do relatively worse. A uniform tax would yield the same result. If this same total abatement level were achieved by allocating tradeable emission permits on the basis of population, and if the market in permits were perfectly competitive and there were no transaction costs, then China would do much better compared with the non co-operative outcome, but the OECD and the ex-USSR would do worse. Finally, if the full co-operative outcome were achieved by allocating tradeable emission permits on the basis of GNP, the OECD would again come out better and the other countries worse compared with the non co-operative outcome. Notice that in these last cases, trading involves making *de facto* side payments. However, the design of these side payments fails to attract broad support.

Without appropriate side payments, agreement among all three parties will prove elusive because each negotiating instrument leaves at least one party (actually, in the above case, two parties) worse off, even while total net benefits are greater compared with the non co-operative outcome. However, with appropriate side payments, any of the negotiating instruments could prove acceptable to all three parties. For example, uniform per cent abatement would be acceptable to all three parties if the OECD transferred $1 000 each to the ex-USSR and China. The OECD would then receive $1 850 more than in the non co-operative outcome, the ex-USSR would receive $339 more, and China $178 more.

Table 2.1. **Full Co-operative and Non Co-operative Outcomes for Hypothetical Negotiations Involving the OECD, the former USSR and China**

	OECD	Ex-USSR	China	Total
Full Co-operative Outcome				
Per Cent Abatement	30.8%	30.8%	30.8%	30.8%
Net Benefits	8 422	940	–601	8 761
Uniform Per Capita Allocation	1.2	1.2	1.2	1.2
Net Benefits	1 220	–2 770	10 245	8 761*
Uniform Per GNP Allocation	2.0	2.0	2.0	2.0
Net Benefits	8 496	922	–634	8 761*
Non Co-operative Outcome				
Per Cent Abatement	27.8%	5.8%	0.8%	18.5%
Per Capita Emissions	2.1	3.3	0.5	1.4
Per GNP Emissions	1.6	4.1	18.1	2.4
Net Benefits	4 572	1 601	221	6 394

1. Numbers do not add, due to rounding errors.
Source: Computed based on simulation summarised in Table 1.15.

We have therefore established that if free-riding were not a problem, and economic instruments were employed to meet the global obligations, the full co-operative outcome could be achieved by a suitable reallocation of the total net benefits of co-operation.

Negotiating side payments

While this is a welcome result, there are many reallocations that can make all parties better off compared with the non co-operative outcome. Negotiations will therefore have to select one reallocation that is acceptable to all parties. Where parties are different, choice of a reallocation will be non-trivial. The problem is in fact similar to the negotiation problem discussed involving the OECD countries. Certain negotiating instruments can make all parties better off. However, some parties will prefer one instrument over another, and some will also want greater uniform obligations than others.

This problem is illustrated in Figure 2.1 for negotiations involving two parties. For purposes of illustration, these might be coalitions of OECD and of developing countries (DCs). The horizontal axis represents developing country net benefits, and the vertical axis OECD net benefits. The point labelled NC represents the net benefits received by these parties at the non co-operative outcome. This point is sometimes called the disagreement point or the threat point of negotiations, because it represents the pay-offs which the two parties receive if negotiations fail. Any point northeast of NC represents an improvement in net benefits for at least one party. Hence, successful negotiations must lead the parties to a point northeast of NC.

Figure 2.1. **The bargaining range**

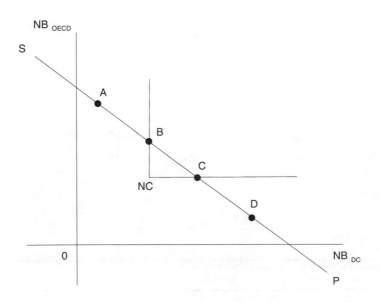

54

The line labelled SP denotes all of the possible ways in which the total net benefits associated with the full co-operative outcome can be divided between the two parties. Any uniform obligation that employs economic instruments – a carbon tax or a tradeable permit scheme – will get countries onto the SP line (for a demonstration, see Barrett, 1991*b*). However, depending on the instrument chosen, and the circumstances of the two parties, one of the parties might be worse off even while total net benefits are maximised. For example, if permits were allocated on the basis of a uniform percent reduction in emissions, it is likely that the outcome would be northwest of point NC but on the SP line – perhaps at point A. If permits were allocated on the basis of population, it is likely that the outcome would be southeast of point NC but on the SP line – say, at point D. In the first case, the OECD countries do better than in the non co-operative outcome, but the developing countries do worse. In the latter, the developing countries do better, and the OECD countries do worse.

The solution will only be acceptable to the two parties if side payments are offered which allocate the net benefits of full co-operation somewhere between points B and C on the SP line. However, there are an infinite number of such points. Determination of the outcome requires an analysis of bargaining.

Allocating the gains to full co-operation

If countries were identical, the solution would likely involve an equal division of the gains from full co-operation. This happens to be the Nash bargaining solution to this "game" (see Barrett, 1991*a*) and the Shapley Value for this game (see below). It is also the outcome that emerges in experimental tests (see Schelling, 1960; and Raiffa, 1982). However, where countries differ, the outcome is, in general, indeterminant.

One might think that a simple rule would suffice to divide the gains from co-operation. For example, one could take the difference between the total net benefits in the full co-operative and non co-operative cases, and allocate this difference among the parties on the basis of their populations. If one were to apply this rule using the figures in Table 2.1, one would obtain the allocation shown in Table 2.2. This rule would naturally favour China. However, the OECD and the former USSR would also prefer this outcome to the non co-operative outcome.

While this rule leaves all three countries better off compared with the non co-operative outcome, the OECD and ex-USSR may argue against the fairness of the rule. The situation would be similar if the allocation were based (inversely) on per capita GNP. Such a rule may seem fair to some, but it need not serve as a focal point. The following example from Raiffa (1982, p. 52) illustrates the point:

> "How should a rich man and a poor man agree to share $200? The rich man could argue for a $150-$50 split in his favour because it would grieve the poor man more to lose $50 than the rich man to lose $150. Of course, an arbitrator, keeping in mind the needs of the rich man and the poor man, might suggest the reverse apportionment. The rich man could also argue for an even split on the grounds that it would be wrong to mix business and charity: 'Why should I be asked to give charity to this poor man? I would rather get my fair share of $100 and give charity to a much poorer person.'"

Table 2.2. **Allocation of Hypothetical Gains from Co-operation on the Basis of Population**

	OECD	Ex-USSR	China	Total
Population (Millions)	851	288	1 136	2 275
Per Cent of Total	37.4%	12.7%	49.9%	100.0%
Allocated Gains	185	301	1 181	2 367
Total Net Benefits	5 457	1 902	1 402	8 761

Source: Population estimates for 1990 computed from World Resources Institute (1990), Table 16.1. The total gains to co-operation are computed from Table 2.1.

The Shapley Value

One argument the OECD and the former USSR may make against the above allocation rule is that it fails to reflect the contributions of the three parties to the generation of the gain in net benefits. Table 2.3 summarizes the net benefits resulting from failure to co-operate, full co-operation and partial co-operation. In the case of partial co-operation, it is assumed that the two co-operating countries seek to maximise their collective net benefits, taking as given the abatement choice of the non co-operating country. The latter country continues to choose its abatement level, taking as given the abatement level of the two co-operating countries. It is these partially co-operative outcomes that tell us something about the contribution which a particular country can make to total net benefits.

One sees from Table 2.3 that it would be credible for the OECD and ex-USSR to threaten not to agree to the per capita allocation rule because, while they would do better with that rule than with no agreement at all, they do better still if they co-operate *independently* of China. If they come to their own agreement, they can earn 7 512 collectively, which is greater than 7 359, the net benefits received by the OECD and former USSR under the per capita allocation rule.

However, if the OECD and ex-USSR co-operate and China does not, total net benefits are 7 822, substantially less than the 8 761 associated with full co-operation. Hence, while it would be credible for the OECD and ex-USSR to threaten not to negotiate on the per capita allocation rule, all three countries would prefer to find an acceptable alternative allocation rule.

One alternative suggested by the game theory literature is the Shapley Value (see Raiffa, 1982; and Friedman, 1986). This value assigns to each country an amount equal to the average of the contributions which that country makes to the total, where the average is taken over all possible sequences through which full co-operation can be formed. The latter is important, for Table 2.3 suggests that the sequence is something on which countries might disagree.

For example, if China joins the coalition consisting of the OECD and USSR, it contributes 8 761 - 7 512 = 1 249 to the total, since 7 512 is what the OECD and ex-USSR would receive if China did not join, and 8 761 is what all three countries receive if

Table 2.3. **Net Benefits to Alternative Co-operative Arrangements**

	Net Benefits
Non Co-operative Outcome	
OECD	4 572
Ex-USSR	1 601
China	221
Total	6 394
Partial Co-operative Outcomes	
OECD, Ex-USSR co-operate	7 512
China does not co-operate	310
Total	7 822
OECD, China co-operate	5 524
Ex-USSR does not co-operate	1 904
Total	7 428
Ex-USSR, China co-operate	1 845
OECD does not co-operate	4 964
Total	6 809
Full Co-operative Outcome	
Total	8 761

all three co-operate. However, if China joins the OECD before the ex-USSR joins, China contributes only 5 524 - 4 964 = 560 to the total, since 4 964 is what the OECD gets if China and the USSR do not co-operate with the OECD, and 5 524 is what the OECD and China get if they co-operate, but the former USSR does not. Finally, if China is the first to form a coalition, it can claim to contribute only 310 to the total, for that is all China receives if it operates independently of the OECD and ex-USSR. China would like to claim that it would be the last to join the coalition. But then so too would the other parties, for they each contribute more to the total when they are the last to join.

Table 2.4 calculates the Shapley Value for the outcomes described in Table 2.3. The OECD would receive 5 764, the former USSR 2 394, and China 603. Compared with the allocation rule in Table 2.2, the OECD and former USSR do better, and China worse. Notice, however, that while the OECD and former USSR can credibly refuse to negotiate on the per capita allocation rule, China cannot credibly refuse to negotiate on the Shapley Value; if it failed to accept the Shapley Value, it would receive 310, or almost half the Shapley Value.

The Shapley Value has much to commend it, and it is a useful device for thinking about the allocation of side payments. But global warming and the global distribution of resources are not issues that can be easily evaluated using such a calculus in practice. Net benefits will be hard to quantify, and subject to uncertainty. Countries may interpret the scientific evidence in different ways, they may have different attitudes toward bearing risk, they may attach different values to potential damages, they may attach different weights to the components that make up net benefits (should those who are harmed by global warming receive the same weight as those who are advantaged?), they may have

Table 2.4. **The Shapley Value to the Coalition Game**

Sequence of Countries Forming the Grand Coalition	OECD	ex-USSR	China	Total
OECD, Ex-USSR, China	4 964	2 548	1 249	8 761
OECD, China, Ex-USSR	4 964	3 237	560	8 761
Ex-USSR, OECD, China	5 608	1 904	1 249	8 761
Ex-USSR, China, OECD	6 916	1 904	–59	8 761
China, OECD, Ex-USSR	5 214	3 237	310	8 761
China, Ex-USSR, OECD	6 916	1 535	310	8 761
Average	5 764	2 394	603	8 761

different rates of time preference, and they may disagree on the ethical basis for using net benefit figures in the first place. Information may be asymmetric; country A may know its net benefits but not those of B, and vice versa. Countries may even have trouble calculating their own expected net benefits, never mind those of other countries. What is more, the number of countries involved would be very large, making calculation of the Shapley Value difficult.

An alternative approach may be to revert to simple rules for effecting side payments – ones that do not require detailed calculation – and ask whether these would be acceptable to the negotiating parties generally. The next section considers a number of rules that have been proposed previously.

SOME PROPOSED SIDE PAYMENT MECHANISMS

Table 2.5 presents a number of mechanisms that have been proposed for making side payments. These proposals do not exhaust the list of possible mechanisms, but an analysis of these should provide some insights into the design of an effective mechanism for global warming.

Donations based on GNP

The first mechanism would involve the OECD countries contributing a fixed percentage of their GNP to an international fund. A proposal using this mechanism was made by Norway in 1989, although the proposal was not supported by other OECD countries. This mechanism has the virtue of being related to ability to pay and, because damages from global warming are likely to be correlated with GNP, to the damages avoided by abating greenhouse gases. Another attraction of this proposal is that it guarantees the level of donations made available. The Norwegian proposal suggested a donation of 0.1 per cent of GNP (see Table 2.5), which would imply transfers of about $11 billion annually from the OECD countries. For comparison, official development assistance by the OECD countries in 1988 ranged from 0.20 per cent of GNP in the case of Ireland (the US donated 0.21 per cent of its GNP that same year) to 1.10 per cent in the

Table 2.5. Some Proposed Side Payment Mechanisms

Rule	Rationale
1. OECD countries contribute 0.1 per cent of their GNP each year to a fund intended to assist developing countries in reducing their emissions.	Based on ability to pay (and, very roughly, the damages avoided by emission reductions). (The proposal was made by Norway at the 15th UNEP Governing Council meeting in 1989.)
2. Choose a global limit on CO_2 emissions, allocate permits to emit CO_2 on the basis of population, and allow permits to be traded internationally. The currency of trading could be either cash or restricted to development project and technology assistance. The price of trading could be either market-determined or fixed administratively.	"It creates a mechanism for ensuring a real transfer of resources from rich countries which are over-exploiting the atmospheric resource to poor countries which need technical assistance if they are to minimise further exploitation as they develop" (Grubb, 1989, p. 38). See also Bertram *et al.* (1989), Hibiki *et al.* (1989), and Agarwal and Narain (1991).
3. Impose an international carbon tax, where the funds are to be collected by an international agency and then redistributed to countries in need of assistance. The tax could be set at a high level to provide an incentive to abate CO_2 emissions, or at a low level solely for the purpose of raising revenue. Alternatively, the tax could be set at a high level by national governments, but with only a small fraction being transferred to an international agency.	Based on the "Polluter Pays Principle". Mechanism for redistributing funds is usually not made explicit. See Nitze (1990).

case of Norway (the next biggest donor was the Netherlands, which donated 0.98 per cent of its GNP; see World Bank, 1990, Table 19).

However, there are two difficulties with the Norwegian proposal. One is that the total amount of money collected is unlikely to be near the amount that would be needed to reach the full co-operative outcome. It could be too high or too low. While much uncertainty surrounds the costs of abatement, a number of studies suggest that the cost of some proposed abatement strategies could range from 2-10 per cent of GNP by 2050 (see Darmstadter, 1991). World GNP is about \$15 940 billion, whereas OECD GNP is about \$11 272 billion[4]. Two percent of the difference amounts to \$93 billion, whereas the Norwegian proposal would provide only about \$11 billion. However, if the full co-operative solution requires only modest reductions in global emissions (as Nordhaus's (1990) analysis suggests), then the costs of abatement will be lower and the discrepancy between the payments corresponding to 0.1 per cent of OECD GNP and the required transfer will be smaller, and could possibly be negative.

The second difficulty with this proposal is that it is silent on the questions of how the money collected would be allocated and what obligations would have to be borne by recipients. This omission does not count against the proposal. However, the proposal is incomplete without specifying a rule for dispersing the funds.

Precedent suggests that recipients will have to be parties to the agreement, and hence will have to undertake the same general obligations as the donor countries. The obligations may be identical, as in the Fur Seal Treaty (all countries agree to cease pelagic sealing), or non-uniform as in the Montreal Protocol (developing countries are allowed to increase their CFC emissions in the short run, and have a longer period in which to achieve the ban). Where the obligations are non-uniform, the main effect will be to limit the magnitude of the side payments; if countries are required to meet an easier target, then the size of the side payment required to secure their agreement will be lower.

Chapter 1 demonstrated that countries will prefer to employ economic instruments in a global warming Convention. If recipients of side payments are required to bear the same obligations as donors, then this means that the Norwegian proposal would probably form part of a global warming Framework Convention involving either tradeable permits or a carbon tax. Hence, the proposal to fix donations can be seen to be related to the other two proposals.

Tradeable permits

The second proposal involves allocating tradeable emission permits on the basis of population, and allowing these to be traded internationally. As noted earlier, this proposal has a built-in mechanism for making side payments. By definition, countries with high emissions per capita receive "too few" permits and countries with low emissions per capita "too many" in the sense that the former would purchase permits from (i.e., provide side payments to) the latter.

Supporters of this proposal argue that its virtue lies partly in the fact that it would transfer resources from rich countries to poor. However, countries with high emissions per capita are not always rich. The ex-USSR emits about 3.5 metric tons of carbon per capita, well above the OECD average of 2.9, and yet the ex-USSR can hardly be classified as "rich." If emissions from deforestation are included, Brazil's emissions per capita could exceed 8, whereas the OECD average would remain 2.9. The top 50 emitters of greenhouse gases on a per capita basis include all the OECD countries except Turkey, but just as many non-OECD countries (World Resources Institute, 1990, Table 3). Indeed, the top emitter is the Lao People's Democratic Republic, which is ranked by the World Bank (1990, Table 1) as the world's 10th poorest country. Proponents of this proposal may limit their attention to CO_2 emissions from fossil fuel burning, but negotiators representing countries that would be disadvantaged under this proposal are more likely to contend that the Convention should limit emissions arising from deforestation (and possibly other emissions) as well.

Unlike the previous proposal, this one is at least complete; given a total emission reduction, the allocation rule combined with the trading mechanism determines the amounts to be paid by donors and the amounts to be received by recipients. The great problem with this proposal is that the transfers involved would only by chance come close to the amounts which donors would be willing to pay and receivers would be willing to accept.

The problem is illustrated in Figure 2.2. If a total abatement (i.e., emission) level has been decided, and emission rights are allocated initially on the basis of population (say allocation A_0 in the figure), then the OECD countries, by virtue of their high emissions per capita, would bear a greater marginal cost of achieving these original obligations. Indeed, it is likely that many poor countries would not have to undertake any abatement at all to comply with the emission per capita ceiling. If trading were permitted, and carried out to the point where all gains from trade had been exhausted, then the final allocation of permits would rest at A^*.

Figure 2.2 **Illustration of potential gains from trade**

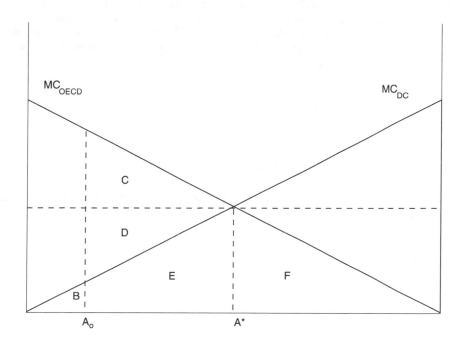

In the absence of trading, the OECD countries could comply with the obligation by incurring a cost equal in area to C+D+E+F, whereas the developing countries would incur a cost equal to area B. With trading, the OECD countries save C+D+E in abatement costs, but pay the developing countries an amount equal to D+E. Hence, the gain to the OECD countries from trading is area C in the Figure. In undertaking the additional abatement, developing countries increase their abatement costs from B to B+E. However, these countries receive area D+E from the OECD, and hence, gain area D from trading.

The developing countries will clearly accept this proposal as long as D>B, even if these countries receive no benefit from abatement. The reason is that the developing countries would then be paid an amount (D+E), which exceeds the actual cost of abatement (B+E). The developing countries would receive a windfall. Countries with a level of emissions per capita below the allocated amount would not need to undertake abatement in the absence of trading, and hence, can only gain from the proposal, even if they receive no benefit at all from global abatement.

The OECD countries will accept this proposal only if the total benefit they receive from the additional abatement exceeds their total cost (D+E+F), *and* if this net benefit exceeds the level associated with the non co-operative outcome (ignoring incentives to form coalitions). Notice that since the proposal does not involve benefits, there is no guarantee that the proposal will ensure that even the former requirement will be met, never mind the latter.

Perhaps recognising that this proposal may mean that donors pay more than they would be willing to, and receivers get more than they would be willing to accept, proponents of this proposal have suggested ways of limiting the magnitude of the transfers. Grubb (1989), for example, suggests that the donors should provide development and technological assistance, rather than cash. However, changing the currency does little to change the magnitude of the cost. What is more, the inefficiency of barter will lower the total gains from trading. Agarwal and Narain (1991) suggest that the price of permits might be pegged at some level. The difficulty here is that a prescribed price is unlikely to clear the market; at least some of the gains from trade will again be sacrificed.

The only effective way of making the proposal more acceptable is to reduce the *total* amount transferred without affecting the *price* at which these transfers take place. The obvious suggestion would be to choose an allocation somewhat to the right of A_0 – say an allocation which ensured that areas B and D were equal. This allocation would guarantee that the developing countries do not lose from the agreement (they gain if they have positive abatement benefits), while the OECD countries would be more likely to come out net gainers.

The Montreal Protocol comes close to eliminating any transfers involved in trading. It establishes quotas based on historical emission levels. Production quotas can be traded internationally. Consumption (defined as production plus imports minus exports) quotas cannot be traded, except within the European Community. In allocating quotas on the basis of historical levels, and restricting trade to production, the Montreal Protocol severely limits the magnitude of transfers. Developing countries are likely to receive no side payments as a consequence of the allocation (i.e., area D in Figure 2.2 is likely to be zero), and as a consequence are unlikely to want to sign the agreement. However, in paying all "incremental" costs, the developed countries ensure that the developing countries do *no worse* by joining (while area D in Figure 2.2 is likely to be zero, so is area B + E)[5]. The Montreal Protocol sacrifices some potential cost savings by not allowing trade in consumption quotas, but it does achieve efficiency in production without huge transfers.

An acceptable trading proposal for global warming would have to share some of these features. The proposal would have to limit transfers resulting from trading by allocating permits on the basis of historical emission levels. To provide an incentive for non-OECD countries to accede to the Convention, a mechanism would also be required to pay "incremental costs." The magnitude of such funds could be unlimited, as in the Montreal Protocol, or determined by a rule similar to the one proposed by Norway.

International carbon tax

Under the third proposal, countries would pay a tax on their carbon emissions, with at least a portion of the revenues being collected by an international agency, which would redistribute the funds on some agreed basis.

Modest versions of this proposal would involve the rich countries setting a very low tax solely for the purpose of raising revenue. The revenue would then be redistributed to poor countries in order to subsidise activities that would reduce net emissions. The European Commission previously considered the idea of a "small" carbon tax, the purpose of which would be to raise funds for protecting tropical rain forests. Although a formal proposal was not made, the Commission calculated that a tax of half a US cent per kwh of fossil fuel use would generate \$55.6 billion annually. Nitze (1990) mentions a tax of just 5 US cents per ton of carbon, but does not say whether this tax has been offered as a serious proposal. These proposals are similar to the proposal to calculate donations on the basis of GNP, in that they all specify an upper limit on donations. However, the carbon tax proposal is consistent with the "Polluter Pays Principle", whereas the GNP proposal is consistent with the "Ability to Pay Principle".

A much more ambitious proposal would involve countries agreeing to an international carbon tax with the revenues being redistributed on the basis of population (this is a variant on the tradeable permit proposal; see Whalley and Wigle, 1990). The simulation presented in Table 2.6, which is based on hypothetical data, illustrates the difficulty with such a redistribution rule. The rule imposes a severe penalty on the OECD and former USSR, which have fairly high emissions per capita, and provides China with a huge windfall.

The OECD could do no better than to have a uniform tax equal to 8.0. With this tax, global abatement is higher compared to the non co-operative outcome (22.9 per cent versus 18.5 per cent) and OECD abatement is lower (22.9 per cent, compared with 27.8 per cent). However, the OECD would have to make net transfers equal to 5 951, and hence receive a net benefit of only 1 402. It is because of the magnitude of the transfer that the OECD countries do worse under this proposal. Under this same tax, the ex-USSR does worse than with no agreement at all, even if transfers are not made, because the proposal imposes such huge abatement costs on the ex-USSR. China receives 40 times the net benefits corresponding to the non co-operative outcome. It is because China benefits so much from this rule that China would prefer an even higher carbon tax. However, such a tax would clearly not be acceptable to the OECD and ex-USSR.

Table 2.6. **Simulation Results for Uniform Carbon Tax With Revenues Redistributed on Basis of Population**[1]

	OECD	ex-USSR	China	Total
Full Co-operative Outcome				
Per Cent Abatement	30.8%	30.8%	30.8%	30.8%
Net Benefits	8 422	940	−601	8 761
Non Co-operative Outcome				
Per Cent Abatement	27.8%	5.8%	0.8%	18.5%
Net Benefits	4 572	1 601	221	6 394
Uniform Carbon Tax				
Preferred Choice	8.0	–	14.9	–
Actual Abatement @ 8.0	22.9%	22.9%	22.9%	22.9%
Net Benefits @ 8.0	1 402	−1 999	8 778	8 181
Net Transfers @ 8.0	−5 951	−3 084	9 034	–
Actual Abatement @14.9	42.5%	42.5%	42.5%	42.5%
Net Benefits @ 14.9	368	−4 036	11 161	7 493
Net Transfers @ 14.9	−8 242	−4 271	12 513	–

1. This simulation is consistent with the one presented in Table 1.15.

A similar story emerges from Whalley and Wigle's (1990) analysis of a carbon tax. Whalley and Wigle choose two types of carbon tax designed to reduce CO_2 emissions 50 per cent compared with the emissions that would be predicted to occur in the absence of any abatement over the period 1990- 2030. The first tax is imposed at the point of consumption, with the revenues being retained by the national governments. The second is also imposed at the point of consumption, but with the revenues being redistributed on the basis of population. Table 2.7 shows that the OECD countries suffer badly under the international carbon tax, whereas the developing and centrally planned economies do better compared with the national tax. Interestingly, the results indicate that the transfers impose a cost far greater than the amount transferred. The total cost net of transfers of achieving the same target using the same instrument varies by a factor of two, simply because of the redistribution[6]. The analyses presented previously in this chapter assume that the cost to a country of making side payments is exactly equal to the magnitude of the side payments. Whalley and Wigle's work suggests that the cost could be much greater.

Part of the problem with the proposal of redistributing revenues on the basis of population lies with the rule for reallocating revenues. An alternative would be to reallocate the revenues to pay for the costs of abatement. However, there is no reason to suppose that the revenue collected by the tax would be equal to the costs incurred. Figure 2.3 shows that the tax would impose a total cost on all countries equal to area A but raise an amount of revenue equal to area B. As long as the tax rate were fairly low, the revenue raised would far exceed the associated costs[7].

Nitze (1990) offers a proposal that attempts to limit this problem. His proposal is that all countries impose a carbon tax equal to $15 per ton. Non-OECD countries would retain their own revenues. OECD countries would retain 80 per cent of theirs, but donate the remaining 20 per cent to an international fund. The fund, perhaps administered by the World Bank (or a combination of the World Bank, the UNDP and the regional development banks), would then redistribute the revenues to developing countries by subsidising projects that lower the costs to these countries of abating their emissions. This proposal has the advantage of limiting the amount of transfers. Nitze estimates that the OECD countries would contribute about $8.25 billion annually, which is even less than the

Table 2.7. **Incidence of National Versus International Carbon Taxes**

	OECD	Oil Exporters	Others	Total
National Carbon Tax				
Cost[1]	–1 110	4 375	5 948	9 213
International Carbon Tax				
Cost[1]	17 473	3 414	–2 371	18 516
Net Transfer[2]	–13 545	1 513	12 032	–

1. Calculated as the Hicksian-equivalent variation measure of welfare change over the period 1990-2 030 in billions of US (1990) dollars.
2. Calculated in billions of US (1990) dollars over the period 1990-2030.
Source: Compiled from Whalley and Wigle (1990), Tables 7 and 8.

Figure 2.3. **Illustration of relative magnitudes of costs
and revenues under a carbon tax**

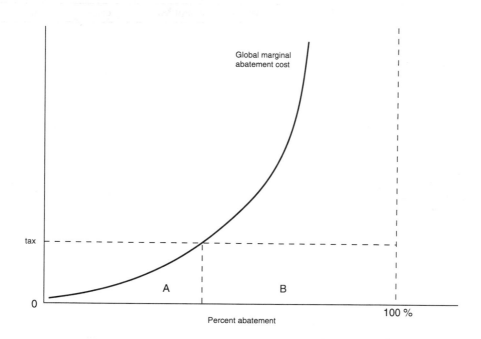

amount that would be transferred under the Norwegian proposal. The proposal would also result in cost-effective abatement, since all signatories to this agreement would impose the same carbon tax[8]. However, it is likely that the funds would have to be rationed, and hence that the number of non- OECD signatories would be limited. In this sense, the side payment mechanism would be similar to that employed by the World Heritage Convention.

To explore this issue a little further, Table 2.8 presents the results of a simulation analysis that is consistent with the analysis presented in Table 2.6. The difference is that the analysis in Table 2.8 assumes that only the OECD countries donate tax revenues to an international fund, and that these donations amount to only 20 per cent of the amount collected. It is further assumed, somewhat arbitrarily, that these revenues are redistributed to the ex-USSR and China in proportion to each country's abatement costs relative to total non-OECD (i.e. the former USSR plus China) abatement costs.

The simulation suggests that Nitze's proposal would be acceptable to the OECD, since the OECD receives higher net benefits with a tax equal to 11.6 compared with the non co-operative outcome. The ex-USSR and China do much better compared with the non co-operative outcome because of the magnitude of the transfers. If more countries were included in the analysis, the OECD would do better still, because total abatement would be increased while OECD transfers would remain largely unchanged. The ex-USSR and China would do worse because the revenues would be shared with more non-

Table 2.8. **Simulation Results for Uniform Carbon Tax With 20% of OECD Revenues Being Redistributed to the former USSR and China**[1]

	OECD	ex-USSR	China	Total
Full Co-operative Outcome				
Per Cent Abatement	30.8%	30.8%	30.8%	30.8%
Net Benefits	8 422	940	−601	8 761
Non Co-operative Outcome				
Per Cent Abatement	27.8%	5.8%	0.8%	18.5%
Net Benefits	4 572	1 601	221	6 394
Uniform Per Cent Abatement				
Preferred Choice	11.6	11.2	7.9	–
Actual Abatement @ 11.6	33.1%	33.1%	33.1%	33.1%
Net Benefits @ 11.6	4 728	3 318	666	8 711
Net Transfers @ 11.6	−3 864	2 471	1 393	–
Actual Abatement @ 11.2	32.0%	32.0%	32.0%	32.0%
Net Benefits @ 11.2	4 722	3 322	704	8 748
Net Transfers @ 11.2	−3 794	2 427	1 368	–
Actual Abatement @ 7.9	22.6%	22.6%	22.6%	22.6%
Net Benefits @ 7.9	4 246	3 033	854	8 133
Net Transfers @ 7.9	−3 047	1 949	1 098	–

1. This simulation is consistent with the one presented in Chapter 1, Table 1.15. Revenues are allocated to the former USSR and China on the basis of each country's abatement cost as a proportion of total non-OECD abatement cost.

OECD countries. However, the ex-USSR and China may still do better compared with the non- co-operative outcome because they are being "over compensated" in Table 2.8. While these results must be viewed with caution, they suggest that Nitze's (1990) proposal would have some appeal.

A variant of this proposal would be for the OECD countries to agree to pay "incremental costs" to all non-OECD signatories, provided the non-OECD countries agree to impose the designated carbon tax. This proposal would not limit OECD contributions, but nor would it restrict participation in the Convention. OECD contributions could then be apportioned on the basis of past or current emission levels, or on the basis of GNP. This proposal would more closely resemble the Montreal Protocol model. Efficient abatement is achieved without transfers, but OECD countries pay non-OECD countries' 'incremental' costs in order to provide incentives for the latter to participate.

USING NEGOTIATIONS ON OTHER INTERNATIONAL ISSUES AS SIDE PAYMENTS

We have so far considered potential climate change as an isolated issue. Broad agreement is necessary if anything substantial is to be done about this problem, but to win broad agreement will require side payments.

Now imagine that a parallel issue is also being treated in isolation. Again, broad agreement is necessary, and to win broad agreement will require side payments. However, suppose that the countries that would be willing to make side payments to increase global abatement of greenhouse gas emissions are the same as the ones that would have to receive side payments if this second issue were to be resolved. Then it is clear that linking the two issues could reduce the need for monetary side payments.

The value of linking issues is forcefully motivated by Sebenius's (1991, pp. 76-77) account of the Law of the Sea (LOS) negotiations:

"... *two separate* negotiations were attempted; until linked, each proved fruitless. With deep-seabed resources the 'common heritage of mankind,' the 'Seabeds Committee' undertook a negotiation over the regime for seabed mining. Developing countries wanted this Convention to offer meaningful participation in deep-seabed mining and sharing of its benefits. Yet the developed countries whose companies potentially had the technology, the capital, and the managerial capacity ultimately to mine the seabed saw no reason to be forthcoming, and these negotiations went nowhere. At about the same time, strenuous efforts by the United States, the Soviet Union, and other maritime powers – that were greatly concerned about increasing numbers of claims by coastal, straits, island, and archipelagic states to territory in the oceans – sought to organise a set of negotiations that would lead to a halt in such 'creeping jurisdiction'. In effect, the maritime powers were asking coastal states, without compensation, to cease a valuable activity (claiming additional ocean territory). Not surprisingly, these discussions over limits on seaward territorial expansion in the ocean yielded scant results."

"Seen as separate 'protocols', these two issues taken independently were not susceptible to agreement. Yet – together with concerns over the living resources and outer continental shelf hydrocarbons – it was ultimately the linkage of these two issues, navigation and nodules, in a bargaining sense that came to be at the heart of the comprehensive LOS conference negotiations."

Sebenius cautions against linkages that are too broadly based, but the idea of linking issues will undoubtedly prove irresistible to climate change negotiators.

One potentially fruitful linkage is with the conservation of tropical rain forests and biological diversity[9]. Protection of these latter environments would yield global climate benefits. But protection would also yield more certain and immediate benefits, since these resources are a part of the earth's "common heritage". Hence, the rich countries should be more willing to offer payments to protect these resources than to simply reduce greenhouse gas emissions.

CONCLUSIONS

Side payments will be needed if an effective global warming Framework Convention is to be successfully negotiated. The reason is that an effective treaty must involve a substantial number of countries with different characteristics – different cost curves, different benefit curves, different per capita emission levels, and so on. As long as obligations in a Convention are fairly uniform, these differences imply that some countries will gain and some lose. The losers will not sign the Convention unless compensated by the gainers. Introducing non-uniform obligations can reduce the magnitude of side payments, but it is unlikely to eliminate the need for side payments in an effective Framework Convention.

It must be emphasized that side payments are not a "gift". The gainers only gain if the losers also sign the Convention. But, by definition, countries will sign a Convention only if they gain. Hence, side payments are best viewed as a mechanism for dividing the *total* gains to co-operation in such a way that every party to the agreement gains, and hence is willing to sign the agreement in the first instance.

There are many rules that might be employed to distribute the total gains to a co-operative agreement. Many of these may appear ethically attractive. However, from a bargaining point of view, the allocation formula should reflect the contributions which countries make to the total gains to co-operation. If it does not, then some countries may do better by forming a coalition and remaining outside the agreement. Hence, an effective Convention should distribute side payments in a way that reflects bargaining strength, even if other considerations, such as ethical views, are also embodied in the side payment mechanism.

One suggested mechanism is a donation by OECD countries to an international fund, where the magnitude of donations is a fixed percentage of GNP. This proposal ensures that the magnitude of transfers is predictable. But the amount of funds raised by such a mechanism may bear little relation to the amount needed to secure an effective agreement. Moreover, the mechanism does not indicate how the funds would be allocated, or what responsibilities recipients would be required to bear. These decisions would presumably have to be made by a committee organised under the Convention.

A system of internationally-tradeable carbon permits would allow the market to allocate funds (where the total amount allocated would depend on the total quantity of permits and the amounts initially allocated to individual countries), and would itself specify individual country obligations. However, tradeable permits are a crude device for effecting side payments. Simple formulas for initially allocating permits may imply that donor countries have to donate more than they would be willing to, even while recipient countries receive more than they would be willing to accept.

Proposals for redistributing revenues raised by an international carbon tax suffer similar problems. If all signatories imposed the same tax, abatement by signatories would be cost-effective. However, the mechanism for redistributing revenues may mean that some countries are required to donate more than they would be willing to, even while some receive more than they would need in order to benefit by signing the agreement.

Formulas can be derived which make both internationally-tradeable carbon permits and an international carbon tax more acceptable to donors. However, it seems likely that simple formulas will result in some countries not signing the Convention, either because the transfers would be too large in relation to the benefits potential donors would receive by signing, or because the transfers would be too small in relation to the costs potential recipients must bear by signing.

To win broad agreement, it may be necessary to supplement the use of economic instruments (or, indeed, command and control) with a separate side payment mechanism. For example, tradeable permits might be allocated in such a way that transfers would be relatively small, but supplemented with a fund that ensured important poor countries had an incentive to sign the agreement. This is basically the approach employed in the (amended) Montreal Protocol. Similarly, an obligation to impose a national carbon tax could be supplemented by a fund which distributed payments to important poor countries to ensure they had an incentive to accede to this Convention. Fund sources could be based on GNP, or a per cent of each nation's carbon tax revenues, or on some other basis.

Despite the advantages to all parties of side payments, some donors may be reluctant to make the sizeable payments that may be required to achieve substantial global reductions in greenhouse gas emissions. This reluctance may be linked to the uncertainty associated with abatement benefits, or to the long lag between the time abatement is undertaken and the time the effects of the abatement in terms of reduced climate change are felt. Linking global warming Convention negotiations to negotiations on a biodiversity convention may prove helpful in overcoming this reluctance.

Notes

1. For details of this Treaty, see Lyster (1985).

2. For a discussion of this Treaty and its impact, see Lyster (1985).

3. Calculation of these costs may prove contentious. For example, a study of the costs of substitution in India assumed that CFCs would become unavailable in 2006 under the terms of the Protocol. However, the costs of substitution would be significantly lower if the existing stock of CFCs were recycled (as the Protocol not only allows, but effectively encourages). Calculating the costs the latter way reduces the total bill by an order of magnitude (Markandya, 1991).

4. These figures were calculated from data on GNP given in World Resources Institute (1990), Table 15.1.

5. In fact, the developing countries would do better by joining because *(i)* they would benefit from the additional abatement; and *(ii)* if they did *not* join, they would be subject to trade restrictions; and *(iii)* they would likely suffer a competitive disadvantage by not employing the more advanced technologies associated with the CFC substitutes.

6. This result is not explained by Whalley and Wigle (1990), and is somewhat surprising. While we might expect that the countries making positive net transfers would suffer a cost in excess of the transfers themlselves, Whalley and Wigle's results indicate that the countries *receiving* net transfers also suffer a cost, net of the transfers, that is higher than in the case where there were no transfers.

7. For a demonstration of this, see Whalley and Wigle (1990). Even with a tax more than five times the price of energy, the revenues raised by a carbon tax (on consumption) exceed the cost by a factor of five (these data come from Whalley and Wigle, Tables 7 and 8).

8. As noted in Chapter 1, an international agreement on a carbon tax would have to establish a benchmark price for fossil fuels (and forestry activities, if emisssions from the latter are included in the agreement). The most obvious benchmark would be prices that reflect all social opportunity costs, excluding those associated with global warming. If this benchmark is employed, then a uniform carbon tax will be cost-effective.

9. Sebenius (1991) suggests that global warming negotiations might be linked to assistance in reducing growth or in resolving more regional environmental problems, such as desertification.

References

AGARWAL, A. and Narain, S. (1991). *Global Warming in an Unequal World: A Case of Environmental Colonialism,* New Delhi: Centre for Science and Environment.

BARRETT, S. (1990). "The Problem of Global Environmental Protection," *Oxford Review of Economic Policy,* 6: 68-79.

———, S. (1991*a*). "International Environmental Agreements as Games," in R. Pethig (ed.). *Conflicts and Cooperation in Managing Environmental Resources,* Berlin: Sprenger-Verlag (forthcoming).

———, S. (1991*b*). "Economic Analysis of International Environmental Agreements: Lessons for a Global Warming Treaty," in OECD, *Responding to Climate Change: Selected Economic Issues,* Paris: OECD.

DARMSTADTER, J. (1991). *The Economic Cost of CO_2 Mitigation: A Review of Estimates for Selected World Regions,* Discussion Paper ENR91-06, Washington, DC: Resources for the Future.

FRIEDMAN, J.W. (1986). *Game Theory with Applications to Economics,* Oxford: Oxford University Press.

GRUBB, M. (1989). *The Greenhouse Effect: Negotiating Targets,* London: Royal Institute for International Affairs.

LYSTER, S. (1985). *International Wildlife Law,* Cambridge: Grotius Publications.

MARKANDYA, A. (1991). "Economics and the Ozone Layer," in D. Pearce (ed.). *Blueprint 2: Greening the World Economy,* London: Earthscan.

MORISSETTE, P.M. and A.J. Plantinga (1990). *How the CO_2 Issue is Viewed in Different Countries,* Discussion Paper ENR91-03, Washington, DC: Resources for the Future.

NITZE, W.A. (1990). *The Greenhouse Effect: Formulating a Convention,* London: Royal Institute for International Affairs.

NORDHAUS, W.D. (1990). *To Slow or Not to Slow: The Economics of the Greenhouse Effect,* mimeo, Department of Economics, Yale University.

PATERSON, D.G. and J. Wilen (1977). "Depletion and Diplomacy: The North Pacific Seal Hunt, 1886- 1910," in *Research in Economic History,* 2:81-139.

SEBENIUS, J.K. (1991). "Crafting a Winning Coalition: Negotiating a Regime to Control Global Warming," in *Greenhouse Warming: Negotiating a Global Regime,* Washington, DC: World Resources Institute.

WHALLEY, J. and Wigle, R. (1990). *The International Incidence of Carbon Taxes,* Paper Prepared for the Economic Policy Responses to Global Warming Conference, Rome, October 4-6.

WORLD BANK (1990). *World Development Report 1990,* Oxford: Oxford University Press.

WORLD RESOURCES INSTITUTE (1990). *World Resources 1990-91,* Oxford: Oxford University Press.

Chapter 3

Free-rider deterrence
in a global warming convention

INTRODUCTION AND SUMMARY

The free-rider problem is often illustrated using the famous "prisoners' dilemma" game. Figure 3.1 provides an example. There are two countries. Each must choose whether to abate or not to abate its emissions. The matrix indicates the pay-offs (net benefits) received by the two countries. If both countries abate their emissions, they each receive 5 (only relative magnitudes matter). If A abates but B pollutes, A gets a pay-off of 2, while B gets 6. If B abates, but A pollutes, the situation is reversed: A gets 6, while B gets 2. Finally, if both countries pollute, they each get a pay-off of 3. It is clear that A would prefer the outcome where A pollutes while B abates; A would receive 6, which is the highest pay-off attainable by A in this game. Similarly, B would prefer the outcome where A abates while B pollutes. In other words, both countries would prefer to free-ride – to receive the benefit of the other country's abatement without having to incur abatement costs itself.

The equilibrium to this game can be found by noting that each country would prefer to pollute, whatever the other country does. If B abates, A prefers to pollute. If B pollutes, A still prefers to pollute. The same is true for B. Hence, both countries will choose to pollute. The paradox is that each country would be better off if both chose to abate their emissions. Each would then get 5, which is greater than the 3 they each get when they both pollute.

One obtains the same outcome in the asymmetric prisoners' dilemma game, depicted in Figure 3.2. In this case, country B would choose to abate its emissions, provided country A abated its emissions. However, country A will pollute, whatever B does, and hence the (Nash) equilibrium for this game still involves both countries polluting, even though both would be better off if they both abated their emissions.

The free-rider problem is sometimes expressed as an incentive to cheat on an agreement. In the above examples, the two countries might have met before the game was played to agree on the actions they should each choose. They might have agreed to abate their emissions, knowing that they would both be better off than if they continued to pollute. However, when the time came for the countries to choose their actions, it is likely that each country would still choose not to abate its emissions, and hence cheat on the agreement. Notice, however, that the same outcome would prevail if the countries did not reach agreement, perhaps because they know that they would not want to carry out their pledge to abate. Hence, free-riding incentives serve to make countries either to want to

Figure 3.1. **Prisoners' dilemma game**

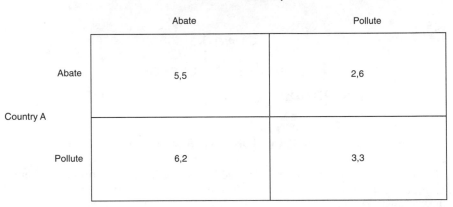

Country B

	Abate	Pollute
Abate	5,5	2,6
Pollute	6,2	3,3

Country A

Payoffs to (Country A, Country B)

Figure 3.2. **Asymmetric prisoners' dilemma game**

Country B

	Abate	Pollute
Abate	5,5	2,4
Pollute	6,2	3,3

Country A

Payoffs to (Country A, Country B)

cheat on an agreement, or to fail to reach agreement in the first place. In richer games involving more than two players, free-riding incentives can lead to incomplete treaties where some countries co-operate and others free-ride. Free-rider deterrence thus makes agreement to abate emissions more sustainable (by making it unattractive for signatories to renege on their commitments), and discourages countries from not signing the Convention in the first place. No agreement to abate greenhouse gas emissions can claim to be successful if it fails to deter free-riding. The purpose of this Chapter is to explore how free-riding might be deterred in a global warming Convention.

Notice that the outcome where both countries abate their emissions (the Pareto optimal outcome) can be sustained as an equilibrium outcome if the parties can make binding commitments. This is the most obvious way of deterring free-riding. Unfortunately, we shall see that binding commitments cannot be made in the case of global warming. There is no trivial solution to the free-rider problem. Alternative mechanisms will be needed if free-riding is to be deterred.

While the prisoners' dilemma game captures the essence of the problem facing negotiators to a global warming Convention, the global warming problem is quite different in nature from the prisoners' dilemma. Choices in the prisoners' dilemma game are discrete; countries choose to abate or not to abate. In the global warming game, choices are continuous; countries choose a level of emissions to abate ranging from 0-100 per cent. In the symmetric prisoners' dilemma game, the (Nash) equilibrium strategies are dominant; each player wants to pollute, whatever the other player does. In the global warming game, each country's preferred abatement level will depend on the abatement choices made by other countries. The options available to the players in the prisoners' dilemma game are twofold: to abate or to pollute. In the global warming game, it might be possible for countries to make a move before the prisoners' dilemma game is played, thereby altering the pay-offs in the prisoners' dilemma game. The prisoners' dilemma game is played once; the players never meet again. The global warming game is played repeatedly. Finally, the prisoners' dilemma game involves just two countries. The global warming game is played by all countries.

The fact that choices are continuous means that the outcome without co-operation need not mean that no action is taken, even if the action that is taken is not Pareto optimal. Failure to co-operate may impose a very slight penalty if countries have a strong incentive to abate their emissions whatever other countries do, or if they would have little incentive to abate their emissions, even if they succeeded in co-operating.

An outcome where little or no abatement is undertaken can also be avoided by the fact that equilibrium strategies in the global warming game are not dominant. (In the asymmetric prisoners' dilemma game, country B does not have a dominant strategy. However, A does, and that is sufficient to foil attempts to co-operate.) If one country reduces its abatement, it may pay the others to increase their abatement. The reason is that reduced abatement by one country increases the marginal benefit of abatement for other countries (assuming that the marginal benefit of abatement decreases as global abatement increases). While the other countries may not make up for the entire reduction in abatement, their response may serve to diminish the harm done by the country that reduced its abatement in the first place.

Taken together, the fact that choices are continuous and that the equilibrium strategies are not dominant means that failure to deter free-riding does not imply that the planet is doomed to suffer an outcome where nothing is done to diminish the greenhouse effect.

The alternative to agreement need not be the "business as usual" scenario used in so many analyses of global warming [see Barrett (1990) for an illustration in the case of ozone depletion]. Still, while something is likely to be done about global warming, it is almost certain that not enough will be done, in the sense that unilateral action will not lead to the full co-operative (Pareto optimal) outcome. Mechanisms to deter free-riding can help. Depending on the nature of the problem, such mechanisms can be critically important.

Recognising the incentive problem in the prisoners' dilemma game, one or both players may want to take actions before the game is played (if that is possible), which alter the underlying incentives. For example, in the asymmetric prisoners' dilemma game, country A knows that B will abate if A abates. A also knows that it does better when both parties abate than when both pollute. Hence, if A can take a decision before the game is played that makes A want to abate when it comes time to play the original game, then A can secure for itself a higher pay-off. The obvious way of doing this would be to make polluting more costly to A. This might be accomplished by investing in pollution abatement capital that makes it more costly to pollute (this strategy only works if the investment is sunk).

If the prisoners' dilemma game is played repeatedly, the free-rider problem *can* be overcome, although this outcome is by no means assured. If countries always meet again, or always have a positive probability of meeting again, then failure to co-operate on one meeting can be punished in the next meeting. While the strategy "cheat on every move" is one equilibrium strategy for this game, it is not a dominant strategy. To see this, suppose the other player adopts the so-called "grim" or "trigger" strategy: "abate on the first move and on every successive move provided the other country abated on the previous move; if the other country did not abate on the previous move, then do not abate on any future move." Then it is not obvious that the other country would do best by not abating on the first move. The potential for punishment explains how countries that care only about their own net benefits can co-operate to attain Pareto optimal outcomes, even when binding agreements are not possible. As noted earlier, a truly binding agreement on global warming cannot be reached. Hence, this ability to punish cheating is vital to the success of a global warming Convention.

Punishment strategies, like the "grim" strategy, may be considered as a self-enforcing agreement. The agreement is enforced by the threat to punish cheaters. This threat supports the "grim" strategy equilibrium, provided it is credible (new literature on renegotiation proof equilibria show that strategies like the "grim" strategy may not be credible. Hence, an agreement to play the strategy may not be self-enforcing). Given that one player cheats, the best move for the other player is also to cheat. This equilibrium is more likely to be sustained when the net benefits to cheating are small, cheating is easily detected, and punishment is both severe and swift. A difficulty with the analysis of repeated games is that there generally exist many equilibria. Under certain circumstances, the full co-operative outcome can be supported as a non co-operative outcome (and hence self-enforcing agreements can be reached), but there is no guarantee that such an outcome will emerge.

One of the primary challenges facing negotiators of a global warming Convention will be the design of credible punishment mechanisms that can deter countries from abating less than demanded by the full co-operative outcome (where it is understood that side payments may be needed to ensure that every country is better off at the full co-operative outcome than at the non co-operative outcome. A stricter definition would also require that every coalition of countries be better off at the full co-operative outcome than at any other outcome; see Barrett, 1991*d*).

This challenge is made daunting by the fact that the number of countries that have a significant effect on the long-term success of a Convention is very large (see Barrett, 1991*c*). It is not necessarily true that an increase in the number of players will make co-operative outcomes more difficult to support (see Shapiro, 1989 for an example in the case of an oligopoly). However, the structure of the global warming game is likely to make credible punishment less effective. The reason is that if one country "cheats," the others suffer only marginally. Hence, only small punishment is credible. But the rewards to "cheating" are large when the marginal abatement cost curve is steep. Hence, credible punishment may not be sufficient to deter cheating (see Barrett, 1991*a*).

This Chapter is organised into four sections. The next Section considers the free-rider problem in greater detail. This is followed by a discussion of the difficulties involved in making binding commitments. The final section considers a number of mechanisms that could be used to deter free-riding. These include leadership, matching, voting, strategic behaviour, and a widening of the game. The main conclusions of the Chapter are as follows:

1. Nordhaus's (1990) analysis of the economics of global warming suggests that even if countries co-operated fully, they would not want to substantially abate their emissions. If his analysis is correct, then Barrett's (1991*a*) analysis of international environmental agreements (IEAs) suggests that failure to deter free-riding may not reduce global welfare by as much as many people have claimed. Still, global welfare would be improved if free-riding could be deterred. And if Nordhaus has underestimated the benefits from abatement, as some economists believe, then the gain to free-rider deterrence could be substantial.

2. Two types of free-riding behaviour will plague negotiations. The direct effect of greater abatement by signatories will be reduced abatement by nonsignatories. The reason is that greater abatement by signatories reduces the marginal benefit of abatement for nonsignatories. However, associated with this direct effect is a price effect. In undertaking greater abatement, signatories will reduce the world price for fossil fuels. This will lead nonsignatories to increase their emissions even more. This price effect will be particularly strong if signatory countries then import energy-intensive products from nonsignatories.

3. The direct effect can be deterred by making accession more attractive to nonsignatories, and this Chapter is largely concerned with how that might be achieved. The price effect can be deterred either by influencing the supply of fossil fuels, such that world prices remain unchanged as a consequence of greater conservation by signatories, or by signatories refusing to import energy- intensive products from nonsignatories. Both forms of deterrence will be costly, and the latter will inevitably undermine attempts to dismantle trade barriers.

4. A major problem for free-rider deterrence is making effective punishment – punishment that succeeds in deterring free-riding – credible. However, the success of free-rider deterrence will also depend on the accuracy and speed of monitoring, and on the speed and clarity of punishment. Negotiators should probably limit the scope of the Convention to pollutants and sources that are easily monitored, even if some of the advantages of a more comprehensive approach are lost in the bargain. Punishment mechanisms would have their

maximal effect if they were automatic – that is, if they were triggered by an observed infraction.

5. Because international law respects sovereignty, binding commitments to abate greenhouse gases cannot be made. Countries can legitimately refuse to sign a Convention, or to withdraw from a Convention at a later date. What is more, if a country violates the terms of a Convention, it can be taken to the International Court of Justice, but only if it agrees to have the case heard. Even then, the Court's decision cannot be enforced. Countries do seem to make binding commitments, but these can be explained by an incentive structure which makes such commitments self-enforcing. Such an incentive structure will not exist in the case of a global warming Convention.

6. The primary impediment to effective punishment is credibility. One credible punishment is suggested by the leadership model. Signatories choose abatement levels which maximise their collective net benefits. Nonsignatories respond by choosing abatement levels which maximise their individual net benefits. If a signatory pulls out of the agreement, the remaining signatories have an incentive to reduce their abatement levels in response. In doing so, the remaining signatories harm the country that pulled out of the agreement. If the harm is sufficiently great, the country will prefer not to withdraw from the Convention, or to accede to the treaty if it had not previously done so. Such a mechanism can be implemented by:

   a) ensuring that the Convention does not come into force until signed by a minimum number of countries accounting for a minimum share of global emissions; or

   b) making it known that the obligations of signatories will be strengthened as the number of signatories increases.

7. This last suggestion is best implemented by the "matching" mechanism. Countries do not agree on emission reductions alone, but on a "flat" level of emission reduction – a level that is independent of the emission reductions promised by other signatories – and a "matching rate" which specifies the additional abatement which the signatory will carry out as a function of the flat abatement levels pledged by the other signatories. This mechanism works by increasing the incentive for countries to join the Convention. If a country joins, the other signatories increase their abatement levels in response.

8. Another possibility suggested by the literature is a voting mechanism, where countries agree to abate their emissions by a given amount provided a minimum number of other countries agree to do likewise. This model may explain why international treaties specify minimum participation levels, as indicated above, but the voting mechanism is not credible. Countries could not be forced to carrying out the promise indicated by their votes.

9. Strategic behaviour may help reduce free-rider incentives. Such behaviour may include overinvestment in abatement capital in order to reduce incentives to cheat, underinvestment in abatement capital in order to make credible punishment more attractive, or legal measures which raise the penalty associated with cheating. Strategic behaviour can be viewed as irreversible commitments made by negotiating parties before agreement has been reached. However, the analysis of strategic behaviour may also suggest that negotiators may want to determine more than just abatement levels. They may also want to include in the Convention procedures for how those levels are to be achieved.

10. Credible punishment is made difficult by the fact that in punishing cheaters, the punishing countries also punish themselves. Negotiators are likely to search for instruments that harm cheaters more than themselves. The obvious example is that of trade barriers. However, while a widening of the game may serve to deter cheating in a global warming Convention, it may also undermine attempts to achieve wider co-operation on international trade policy.

THE FREE-RIDER PROBLEM

The essence of the problem facing global warming negotiators is that abatement by any one country benefits all countries but imposes a cost only on the abating countr[1]. All countries do best if they all co-operate and abate their emissions substantially[2]. But country A would do better still if the others abated their emissions substantially, and A much less so. The same is true for countries B, C, D, etc. Hence, there exists the potential that all countries might fail to abate their emissions substantially, even though they would all be better off if they did so.

The problem is illustrated in Table 3.1, which is based on the analysis presented in Chapter 1 (Table 1.10). At the full co-operative outcome, total net benefits for the United States, the European Community and Japan are maximised. Abatement is cost-effective; each country faces the same marginal cost of abatement. Abatement is also efficient for this group; the marginal cost of abatement is equal to the sum of the marginal benefits of

Table 3.1. **Free Rider Incentives for Case where $c = 35$, $b = 1$**

	US	EC	Japan	Total
Full Co-operative Outcome				
Per Cent Abatement	25.9%	25.9%	25.9%	25.9%
Net Benefits	1 034	1 094	767	2 895
Marginal Cost	9.1	9.1	9.1	9.1
Marginal Benefit	4.1	3.3	1.7	9.1
Non Co-operative Outcome				
Per Cent Abatement	12.4%	10.0%	5.3%	10.8%
Net Benefits	731	721	438	1 891
Marginal Cost	4.3	3.5	1.8	1.8-4.3
Marginal Benefit	4.3	3.5	1.8	9.6
Uniform Per Cent Abatement				
US Preferred Choice	21.6%	21.6%	21.6%	21.6%
Net Benefits	1 078	1 051	686	2 814
Marginal Cost	7.5	7.5	7.5	7.5
Marginal Benefit	4.1	3.3	1.8	9.2
EC Preferred Choice	27.0%	27.0%	27.0%	27.0%
Net Benefits	1 007	1 096	786	2 889
Marginal Cost	9.5	9.5	9.5	9.5
Marginal Benefit	4.0	3.3	1.7	9.0

abatement. However, each country's marginal cost of abatement exceeds its marginal benefit. The cost of abating the last ton of carbon exceeds the associated benefit. Each country has an incentive to reduce its abatement (holding constant the abatement of other countries), because such a reduction will lower abatement costs much more than abatement benefits, and hence increase the country's total net benefits. All countries face exactly the same incentive, and yet if all reduce their abatement, each does worse compared with the full co-operative outcome.

What abatement levels will countries choose? One equilibrium for this game is the set of abatement choices that satisfy the condition that no country could do better by choosing a different abatement level, provided the other countries do not choose different levels. This equilibrium (known as the Nash equilibrium) is compelling, because it implies that any country, after having observed the abatement levels chosen by the other countries, does not regret the abatement choice that it made[3]. This equilibrium is labelled as the "non co-operative outcome" in Table 3.1. Each country's marginal cost of abatement equals its marginal benefit, and hence no country wants to alter its abatement choice unless other countries alter theirs. However, each country's marginal cost of abatement is less than the total marginal benefit of abatement for the group, and hence net benefits for the group are not maximised.

Notice that while all countries would like to avoid the non co-operative outcome (all countries do better at the full co-operative outcome, and hence this outcome is preferred by every country to the non co-operative outcome), they each undertake some abatement when they fail to co-operate. Thus, it is *not* obvious that co-operation would make a huge difference to the well-being of these countries. Barrett (1991*a*) has shown that the gains to co-operation are likely to be greatest for global environmental problems when the slope of every country's marginal abatement cost curve is close in size (in absolute value) to the slope of the global marginal benefit curve, and both of these slopes are "large" (in absolute value)[4].

Nordhaus's (1990) analysis of the economics of global warming suggests that the slope of each country's marginal abatement cost curve is steep beyond some point, but that the slope of the global marginal abatement benefit curve is not steep, given current scientific knowledge. Hence, coupling Nordhaus's analysis with Barrett's implies that the non co-operative outcome may not be very far from the full co-operative outcome. Countries would, on their own, undertake some inexpensive abatement. But even if they co-operated fully, they would probably not undertake very substantial levels of abatement. Whether this is true is not known for certain. If a threshold were discovered such that beyond this level, increases in emissions increased potential damage substantially, then the global marginal abatement benefit curve would be steep in the neighbourhood of the threshold, and the gains to co-operation would likely be very large indeed. There is concern that such a climatic threshold may exist. However, none has been identified yet. Any knowledge about the potential existence of such a threshold would obviously be of extreme importance to policy.

While the non co-operative outcome may not be disastrous, it is certainly not desirable. Countries would do better if they co-operated. Chapter 1 demonstrated that, for the case examined in Table 1.1, the OECD countries could agree on uniform percentage abatement (without side payments). The US, the EC and Japan would accept the level preferred by the US (21.6 per cent) or the level preferred by the EC (27.0 per cent), or some other level not much greater or less than these, such as the full co-operative outcome (defined over the OECD countries alone). (Japan would prefer a much higher level, but this would not be acceptable to the US; see Barrett, 1991*c.*) These levels are "acceptable" because they make each country better off compared with the non co-operative outcome. However, these levels will be difficult or impossible to sustain. For at all of these levels, each country's marginal cost of abatement exceeds its marginal benefit. Every country thus faces a strong incentive to reduce abatement. Hence, while countries may "agree" on abatement levels, this agreement assumes that commitments are binding. If commitments cannot be made binding – a matter discussed later in the Chapter – then actual agreement may not be possible simply because countries would know that none had a sufficient incentive to comply with the "agreed" obligations; if an agreement were reached, it would be unstable.

The mechanics of free-riding

Consider how free-rider incentives operate in detail. Because countries are interdependent, if one group of countries – say, signatories to a global warming Convention – increase their abatement levels, free-riding behaviour would lead the other countries – the nonsignatories – to actually reduce their abatement in response (assuming all else is constant). The reason is that abatement by signatories reduces the marginal benefit of abatement for nonsignatories. Since nonsignatories continue to choose abatement levels where marginal costs equal marginal benefits, the best response of nonsignatories to greater abatement by signatories is to reduce their abatement. To be sure, this response may be barely noticeable for some countries. Countries with flat marginal abatement benefit schedules (low-income countries that are not vulnerable to potential climate change) would only respond marginally to greater abatement by signatories. However, summing up even these marginal changes over a large number of nonsignatories may still yield a significant total response. This concern about the response of nonsignatories is not a theoretical artefact. The US government expressed this exact concern when questioned why it would not take further unilateral action to abate CFCs (see Barrett, 1990).

In the case of global warming, the magnitude of free-riding will be amplified by price effects set into motion by greater abatement on the part of signatories. To reduce their emissions, signatories will have to reduce their consumption of fossil fuels. As a consequence, world prices for fossil fuels will fall. Non-signatory consumption will then rise, as will nonsignatory emissions. The rise in nonsignatory consumption is likely to be quite large, because as these countries gain a cost advantage in energy-intensive industries, world production in these industries will shift to nonsignatories, and the products of these industries will be imported by signatory countries.

In this Chapter, the former free-rider effect shall be termed the "direct" effect, and the latter the "price" effect. The combined effect of these free-riding incentives poses a formidable challenge to negotiators. How might these incentives be altered?

Deterring the direct effect

Because abatement of greenhouse gas emissions is a global public good, there is nothing that signatories can do to prevent nonsignatories from free-riding as a direct consequence of abatement by signatories. If signatories reduce emissions to improve their own environment, they can't help but improve the environment of nonsignatories as well. However, there may exist indirect means. For example, if the signatory countries develop technologies that allow them to reduce the potential damage associated with climate change – such as drought-resistant seed varieties – then the Convention could ban these from being exported to non-signatories. The effect would be to keep the marginal benefit of abatement higher than it otherwise would be, and to provide an incentive for nonsignatories to accede to the Convention. This provision will itself be difficult to enforce, because the country that develops the improved technology would want to market it widely. What is more, the deliberate withholding of such technologies may be seen to be "unfair", particularly if the countries being deprived of the technologies are poor and the countries that would benefit from the ban are rich. Finally, nonsignatories may themselves develop technologies that can lower the costs of adaptation, and reciprocate by refusing to make these available to signatories. If mechanisms are to be developed which alter the incentives to free-ride, it is likely that they will have to operate by increasing the benefit to accession rather than by increasing the benefit to abatement outside the agreement. Much of this Chapter is devoted to mechanisms that do just that.

Notice that signatories have nothing to gain by withholding technologies from nonsignatories that can lower abatement costs. Such technologies would provide an incentive for nonsignatories to increase their abatement even while remaining outside the agreement. The availability of such technologies would also lower the costs to these countries of acceding to the Convention.

Deterring the price effect

To prevent world fossil fuel prices from falling as a consequence of the Convention obligations, signatories would have to reduce world supply of fossil fuels by an amount equal to the reduction in demand. One suggestion would be to restrict the export of fossil fuels from signatories to nonsignatories. Another, and more ingenious, suggestion would be for signatories to buy up a quantity of fossil fuel reserves, and withdraw these from the market. Bohm (1990) argues that the latter option would be a cheaper way for signatories to reduce global emissions than simply reducing signatory demand for fossil fuels alone.

Even if such actions were taken to leave world fossil fuel prices where they would have been in the absence of signatory abatement, nonsignatories will face a lower relative price for fossil fuels (after tax), and hence have a cost advantage. The danger would be that production would shift to nonsignatory countries, and that the products of these industries would be imported by the signatories. In that case, a portion of the emission reduction achieved through abatement in signatory countries would be offset by emission increases in nonsignatory countries. To prevent this from happening, the Convention would have to include provisions for restricting imports from nonsignatories of products requiring a high degree of fossil fuel inputs. In theory, such restrictions would apply only to production that shifted as a consequence of the abatement actions by signatories. In practice, such production will prove exceedingly difficult to identify. What is more, there may also be a temptation for countries to use trade restrictions as a direct deterrent to free-riding (see Whalley, 1991 – a matter discussed later).

Negotiators may draw some inspiration from precedent. The Montreal Protocol bans imports of CFCs from, and exports of CFCs to, nonsignatories. The treaty also includes provisions for banning imports from nonsignatories of certain products containing CFCs, and to later "determine the feasibility of banning or restricting, from States not party to this Protocol, the import of products produced with, but not containing, controlled substances ..." Hence, the Montreal Protocol would seem to provide a model for drafting a global warming Convention.

However, the global warming problem is very different. The cost of substituting for CFCs is much, much lower. Production is concentrated in only a few firms, and the ones that are developing substitutes have every incentive to want to market these widely; a world wide ban works in their favour. Fossil fuel suppliers do not have the same incentive, and will not gladly support efforts to limit demand for their product. The point can be made rather forcefully by noticing that the major oil exporting countries will have little incentive to join a Convention which reduces the demand for their oil. But if they don't join, will the signatories ban imports of their oil? The United States, Western Europe, and Japan import 42 per cent, 75 per cent, and 98 per cent, respectively, of their oil requirements. An oil import ban would cripple these economies. What is more, since virtually every product is made using fossil fuels, a ban on products made using fossil fuels amounts to a total trade ban, and that too is inconceivable.

Constraints on free-rider deterrence

It has already been argued that credibility is a major constraint on free-rider deterrence. However, it is not the only constraint. The ability to deter free-riding will also depend on the accuracy of measures which detect cheating, and on the speed with which the punishment can be made. Negotiators should be careful to limit obligations to areas where compliance is easily monitored. This may mean giving up some of the advantages of the comprehensive approach to abatement (see Task Force on the Comprehensive Approach to Climate Change, 1991). However, the alternative will be to weaken attempts to enforce the Convention's obligations (for a considered discussion of the verification problem, see Fischer *et al.* 1990). The punishment mechanisms must also be able to quickly respond once non-compliance has been observed. It is thus best that they be automatic. An example is the provision in the Montreal Protocol which restricts trade in CFCs with nonsignatories.

BINDING COMMITMENTS

The most obvious deterrent would be binding commitments made by the negotiating parties. If every country agreed to undertake a level of abatement consistent with the full co-operative outcome, and these commitments could be made binding, then countries would make these commitments, and the full co-operative outcome would be achieved. (If side payments were permitted, the pay-offs received by individual countries at the negotiated outcome may differ from the ones shown in Table 3.1, but the abatement levels and the total net benefit would be the same; see Chapter 2.)

It would seem from precedent that binding commitments can be made. However, we shall see shortly that binding commitments cannot be made. Countries can always get out of their commitments if they want to. Hence, an agreement to control greenhouse gas emissions must be self-enforcing. Perhaps the most famous example of a seemingly binding commitment involving transfrontier pollution is the *Trail Smelter* Arbitration. The case involved pollution emissions from a smelter in Trail, British Columbia, which were believed to be damaging crops downwind in the state of Washington. In 1935, the

United States and Canada signed a convention in which they agreed to refer the matter to binding arbitration. The two governments each chose one member of the arbitration tribunal, and a third member jointly (the third member was Belgian). The convention states that the objective of the agreement is "to reach a solution just to all parties concerned", and hence the intention in signing the convention was to push both parties nearer the full co-operative outcome (or at least a Pareto optimal outcome). This is implied in a statement made by the Tribunal (*Trail Smelter* Arbitral Tribunal, 1941, p. 685):

"As between the two countries involved, each has an equal interest that if a nuisance is proved, the indemnity to damaged parties for proven damage shall be just and adequate and each has also an equal interest that unproven or unwarranted claims shall not be allowed. For, while the United States' interests may now be claimed to be injured by the operations of a Canadian corporation, it is equally possible that at some time in the future Canadian interests might be claimed to be injured by an American corporation."

The fame accorded the *Trail Smelter* case is due to its appearance as an exception to the rule that binding commitments are not made. Most international treaties include only vague and weak provisions for enforcement. As examples, the Sulphur Emissions Protocol to the Convention on Long-Range Transboundary Air Pollution offers the following mechanism for the settlement of disputes:

"If a dispute arises between two or more Parties as to the interpretation or application of the present Protocol, they shall seek a solution by negotiation or by any other method of dispute settlement acceptable to the parties to the dispute."

The Antarctic Treaty is more specific, but no more forceful:

"If any dispute arises between two or more of the Contracting Parties concerning the interpretation or application of the present Treaty, those Contracting Parties shall consult among themselves with a view to having the dispute resolved by negotiation, inquiry, mediation, conciliation, arbitration, judicial settlement or other peaceful means of their own choice."

"Any dispute of this character not so resolved shall, with the consent, in each case, of all parties to the dispute, be referred to the International Court of Justice for settlement; but failure to reach agreement on reference to the International Court shall not absolve parties to the dispute from the responsibility of continuing to resolve it by any of the various peaceful means referred to [above]."

The Montreal Protocol on Substances that Deplete the Ozone Layer deferred the preparation of rules "for treatment of Parties found to be in non-compliance" until the first meeting of parties to the protocol, but at the first meeting, the matter was again deferred until the second meeting.

In most IEAs, consent to settle a dispute by reference to the International Court of Justice or to arbitration must be given by both parties, and such consent is usually given only after a dispute has arisen and the nature of the dispute is known. The weakness of this approach is that one party – usually the "guilty" party – may protest against the action and refuse to have the matter adjudicated.

An exception to this approach of having unanimous consent is the Convention on the Conservation of European Wildlife and Natural Habitats (the Berne Convention), which authorises just one party to a dispute to take the matter to arbitration. The approach is similar to that followed in the *Trail Smelter* case:

"Each party shall designate an arbitrator and the two arbitrators shall designate a third arbitrator. ... if one of the parties has not designated its arbitrator within the three months following the request for arbitration, he shall be designated at the request of the other party by the President of the European Court of Human Rights within a further three months period. The same procedure shall be observed if the arbitrators cannot agree on the choice of the third arbitrator within the three months following the designation of the two first arbitrators."

"... The arbitration tribunal shall draw up its own Rule of Procedure. Its decisions shall be taken by majority vote. Its award shall be final and binding."

Why would countries agree to submit disputes to binding arbitration in advance of the disputes arising? The explanation is hinted at in the *Trail Smelter* Arbitral Tribunal's statement quoted earlier.

Consider an example. Suppose there are just two parties to an agreement. Both parties know that at certain moments in time, one will want to "cheat". If enforcement requires the assent of both parties to a dispute, then the cheater will know that the terms of the agreement cannot be enforced, and hence will cheat. The ability to refuse third party intervention benefits the cheater, since by definition this country does better by cheating. However, at a later date, the other party may want to cheat, and this may harm the party that cheated previously. Each party would prefer to be able to cheat itself, but for the other party not to be able to cheat. Realising that this outcome cannot be sustained, the parties may both wish to agree in advance that cheating by either party will be referred at the request of just one party to arbitration by a third party. If the two parties can be sure that the third party will make decisions that are "just" in the sense that they support the full co-operative outcome, then each party would do better on average (or in present value terms) by agreeing to third party intervention. In fact, if detection is certain, it may never be in the interests of the two countries to cheat; the arbitration mechanism would never be used but it would serve a purpose: it would deter cheating.[6]

While the well-being of states can be enhanced by their willingness to effectively relinquish some sovereignty, this is strongly resisted in practice. One reason for this seems to be the feeling that decisions taken by third party arbitrators are not "just". The United States and Canada submitted their dispute about the Gulf of Maine boundary to the International Court of Justice, but the Court rejected boundary concepts based on fish stocks or geological formations "both because they were too uncertain and because the judges were not competent to evaluate the scientific evidence" (Birnie, 1988, p. 98, emphasis added). Instead, the judges drew boundaries by reference to the geographical configuration of the coastline, a decision that was not seen to be fair. "Unfortunately, the outcome, which some observers saw as an unhappy compromise between the Canadian and US positions, may not encourage either state to repeat the experience" (Springer, 1988, p. 56).

Another problem with this procedure is that a state can always refuse to comply with the International Court of Justice's decision. "In the first case ever brought before the Court, the *Corfu Channel* case of 1948, damages of £ 843 947 were awarded to the U.K. against Albania, but not a penny has ever been paid" (Lyster, 1985, pp. 11-12). Arbitration can really only work when countries are locked into a continuing relationship, such that "cheating" can always be punished at a later date. That is, starting at any arbitrary date (and not just the date at which the treaty is negotiated), it must be the case that the present value benefits to each party of agreeing to third party intervention exceeds the present value costs. This condition may well not have been satisfied in the *Corfu Channel* case. It is likely to have been satisfied in the *Trail Smelter* case. It is the combination of "unjust" decisions and the inability to force a country to comply with a third party decision (including decisions by the International Court of Justice), even if the party previously agreed to have the case adjudicated by a third party, that makes binding commitments virtually impossible. Agreements among countries, if they are to succeed, must be self enforcing. Signatories must face incentives that make them want to do what they agreed to do. In the jargon of game theory, IEAs must satisfy the constraint of individual rationality.

This last constraint is particularly important because countries can always withdraw their commitments. All treaties include provisions for withdrawal, although these provisions usually do not allow for instantaneous withdrawal. For example, the Montreal Protocol does not allow withdrawal until four years after the signatory assumed the principal obligations of the Protocol, and even then the withdrawal is not recognised to have taken effect until at least one year after notification.

Below we consider a number of mechanisms that satisfy the individual rationality constraint and yet also commit countries to doing more than they would do if they withdrew from the Framework Convention, or did not sign the Convention in the first place. One of these (strategic behaviour) actually works by making commitments more binding than they would otherwise be.

ALTERING THE INCENTIVES TO FREE-RIDE

In a self-enforcing Convention, free-rider deterrence must alter the incentives to free-ride by making participation relatively more attractive. This section considers a number of possibilities.

Leadership

One mechanism is suggested by Barrett's (1991*a*) model of an IEA. In the model, signatories choose their abatement levels collectively so as to maximise net benefits for the group of signatories. As noted in Chapters 1 and 2, this full co-operative outcome (where "full" is with respect to signatories alone) can be reached, even if countries differ substantially, provided economic instruments are used to ensure that abatement for all signatories is cost effective and side payments are employed to make sure that each party is better off in the agreement than outside it (this assuming that commitments can be made binding).

In choosing their abatement levels, signatories recognise how nonsignatories will respond. Nonsignatories then take the levels chosen both by signatories and other nonsignatories as given, and choose their abatement levels. One way of thinking about this process is that the signatories act as "leaders". They move first. Signatories can act as leaders by virtue of their size; collectively, they are large compared with any nonsignatory. The abatement levels chosen by nonsignatories correspond to the expectations of the signatories, and hence the outcome is a Nash equilibrium. The stable IEA is one where no signatory would prefer to withdraw from the agreement, and no nonsignatory would prefer to accede to the Convention.

The punishment mechanism that makes countries want to join the IEA works as follows. Each signatory chooses an abatement level where its marginal cost of abatement equals the total net benefit of abatement for all signatories. Each nonsignatory chooses an abatement level at which its marginal cost of abatement equals its marginal benefit of abatement. By definition, if there are two or more signatories, each abates more than in the non co-operative outcome, and hence each would like to withdraw. However, if a country withdraws, the remaining signatories reduce their individual abatement level because now the total marginal benefit of abatement is being summed over one less signatory. The withdrawing country naturally lowers its abatement level, and this has the effect of raising every other country's marginal benefit of abatement, but the former effect always exceeds the latter, and as a consequence, each signatory always abates less after a withdrawal. That is the punishment. If a signatory withdraws, it is able to lower its abatement costs. But as a consequence of withdrawing, the remaining signatories reduce their abatement levels, and this action lowers the withdrawing countries net benefits. Withdrawal only appears attractive if the saving on abatement costs exceeds the loss in abatement benefits.

Similarly, if a country joined the agreement, the total marginal benefit of abatement for the group of signatories would be increased, and each original signatory would undertake more abatement than before the accession. The joining country loses by having to bear higher abatement costs itself. But it gains by benefiting from the greater abatement by the original signatories. Joining appears attractive if the increase in abatement costs is exceeded by the gain in abatement benefits.

Notice that this mechanism is self-enforcing. Signatories can do no better than to maximise their collective net benefits. Hence, their "threat" to reduce abatement upon the withdrawal of a signatory is credible.

One way in which this mechanism might operate in a Framework Convention would be for the obligations of existing signatories to increase by some pre-specified amount upon the accession of a new signatory. Presumably, the increase in abatement by existing signatories would depend on the total reduction in emissions brought about by the accession itself. This explicit mechanism is likely to be expressed in a formula, and is similar to the "matching mechanism" discussed later.

Cruder mechanisms can also be used, and seem to be more common. As an example, the Montreal Protocol did not come into force until at least eleven countries accounting for at least two-thirds of 1986 CFC consumption had ratified the Convention. The reasoning behind this provision may have been that if this requirement had been met, then it would be in the interests of signatories to carry out the obligations described by the Convention. If fewer countries, representing a smaller share of global consumption, had signed the Convention, then it would have paid the signatories to undertake weaker obligations. This mechanism, however, is also like the voting mechanism discussed later.

Another crude mechanism would be for the Convention obligations to be reviewed upon the accession of a certain number of countries. The implication would be that as the number of signatories increased, it would be in their collective interests to increase the obligations. Treaties are reviewed periodically, though often because scientific evidence or economic circumstances change over time. When the Montreal Protocol was renegotiated in June 1990, the obligations were strengthened significantly. This is partly because in the months following the original negotiation (the Protocol was agreed in September 1987), estimates of the potential damage from ozone depletion increased, while estimates of the costs of abatement fell. However, the fact that the number of signatories had increased from 24 to 57 over this period may also have had an influence.[7]

The fundamental difficulty with these mechanisms is that, like all free-rider deterrent mechanisms, they work only if they are credible (see Barrett, 1991b). In the case of global warming, the number of countries involved is all countries. Where each is small, accession by any one can only affect the power of the IEA marginally. But if the effect of accession is marginal, then the credible reward for accession will also be marginal. The difficulty is that the cost of accession is not marginal for the country contemplating accession. Barrett's (1991a) model predicts that where many countries contribute to the pollution, and the slope of the marginal abatement cost curve is large for each of these countries, an IEA will not be signed by many countries, even if the slope of the marginal abatement benefit curve is also large. (To verify the usefulness of the model, notice that the model also predicts that the number of signatories to the Montreal Protocol would be large, and that each would agree to undertake substantial emission reductions.)

Effective free-rider deterrence in a global warming Convention must therefore make the benefit of acceding to the Convention non-marginal. Two mechanisms that serve to do just that are discussed below.

Matching

One way of trying to improve upon the outcomes described thus far is to alter the choice variable or instrument. It has been assumed, both here and in previous Chapters, that countries choose an abatement level (possibly allowing for trading) or a tax. However, other instruments could also be chosen.

Guttman (1978) has shown that "matching" behaviour can move us closer to the full co-operative outcome[8]. Indeed, under certain circumstances, matching can attain the full co-operative outcome. This despite the fact that the "game" that is being played is non co-operative.

As noted earlier, free-rider incentives "bite" hard in the case of global warming because each country must pay a substantial cost to abate its own emissions, and yet the reward that any one country receives for abating its emissions is small. Matching alters this incentive. When one country undertakes more abatement, others undertake more in response. The benefit associated with the one country's abatement is thus increased substantially. As a consequence, every country finds it attractive to increase its abatement above the non co-operative level.

Rather than choose an abatement level directly, countries choose a "flat" level of abatement and a matching rate. Formally, country i's abatement level q_i is determined by:

$$q_i = a_i + b_i \Sigma a_j, \quad j \neq i,$$

where a_i is i's flat abatement level and b_i is i's matching rate. If the b_i are all zero, then we have the same problem analysed earlier. An increase in country i's abatement level does not bring about a direct change in other countries' abatement level (it will, or course, bring about an indirect change; other countries will free-ride on this country and reduce their abatement levels in response). However, if countries have positive matching rates, then an increase in a_i will cause other countries to increase their abatement levels.

The matching game is played in two stages. Each country first considers the flat level of abatement that maximises its net benefits, taking matching rates (and the flat levels of other countries) as given. They then choose a matching rate (taking the matching rates of other countries as given). In choosing a matching rate, each country considers how the choice of a matching rate will influence the choice of flat levels.

It can be shown, using the model outlined in Chapter 1, that if there are N countries and all are identical, each will maximise net benefits by choosing a matching rate equal to one, and a flat rate equal to 1/Nth of the full co-operative abatement level. Even though the game that is played is still non co-operative, the full co-operative outcome is achieved.

Matching is a common phenomenon. The U.S. federal government often matches state expenditure on the provision of certain public goods. Charitable donations are also often contingent on the donations provided by others. However, I am not aware of any matching presently being used in IEAs.

One reason for this is that if countries are different and great in number, matching may fail to attain an efficient (i.e. Pareto optimal) outcome, although even under these circumstances it appears that matching will do better than the non co-operative outcome without matching (see Guttman, 1978). Perhaps more worrying is experimental evidence suggesting that matching may not work well when the number of players is greater than even a few in number: " ... computation and observation of matching rates is costly, and these costs may increase with group size, diminishing the returns to matching behaviour" (Guttman, 1978, p. 255).

Voting

There is one very simple solution to the free-rider problem. Have every country vote on a proposal to impose the full co-operative outcome, with side payments being arranged in such a way that every party is better off compared with the non co-operative outcome. Each country must vote to accept the agreement or not, where it is understood that support for the measure must be unanimous. If just one country votes against the proposal, the measure is not adopted and all countries remain locked forever in the non co-operative outcome. Then each party's contribution is suddenly non-marginal. Every party (if rational) would vote in favour of the measure.

There are, however, two problems with this proposal. First, as noted in Chapter 2, it would not be easy to identify the abatement levels and side payments necessary to ensure that all parties were better off compared with the non co-operative outcome. This is especially so when one considers the possibility of countries forming coalitions of more than one, but less than all, countries. Second, countries may be reluctant to agree to the vote if they think circumstances will change so that a better proposal might later be forthcoming.

Black, Levi and de Meza (1990) consider a model which tries to overcome the first problem. Countries know their own net benefits and the distribution of net benefits from which the actual net benefits of other countries are drawn. However, no country knows exactly the net benefits of other countries. Black, Levi and de Meza conjecture that one would require something less than unanimity, because there would exist a chance that some countries could only be made worse off by voting in favour of the measure, and hence would vote against it. However, they also conjecture that if the number of "yes" votes needed to carry the measure fell too low, then free-rider incentives would become too large. Countries would expect that the minimum number of "yes" votes was far below the minimum needed to carry the measure, and hence would prefer to free-ride by voting against the measure. The trick is to specify a voting rule that requires neither too many nor too few "yes" votes. Black, Levi and de Meza's (1990) paper is devoted to determining the minimum participation level needed to maximise expected social welfare.

The voting mechanism discussed by Black, Levi and de Meza may explain why international treaties specify a minimum participation level. The Montreal Protocol, for example, would not have come into effect until ratified by at least eleven countries, accounting for at least two-thirds of the 1986 level of global CFC consumption. However, as noted previously, this mechanism can also be viewed as a means of implementing the leadership model.

The main reason for believing that minimum participation levels are not the result of a voting mechanism is that voting mechanisms are not credible. Consider the unanimity rule for the case that introduced this topic. Suppose every country but one voted in favour of the measure. Then the countries that agreed to the proposal would have every incentive to put forward another proposal that required a unanimous vote but which excluded the country that voted down the previous proposal. The requirement that countries remain locked forever in the non co-operative outcome would not be credible, because when called upon to act on the threat, countries would have every incentive not to act on it. (Presumably, the country that voted "no" knew this would happen.) Having seen that one country got away with voting down the earlier proposal, others may adopt the same strategy in this second vote. And so on. It would seem difficult to sustain an outcome other than the non co-operative outcome using the voting mechanism. (Technically, the strategy of voting "yes", however the other countries vote, is a Nash equilibrium, but it is not subgame perfect.) Voting can overcome free-riding only if the votes are binding. But if commitments can be made binding, we know that the free-rider problem can be easily eliminated.

Strategic behaviour

One way in which the inefficient outcome of the prisoners' dilemma game can be subverted is if players make strategic moves that alter their own best choices in subsequent moves. For example, a country might want to make an investment which serves the purpose of lowering the cost to punishing defection in subsequent moves. If the investment served to make punishment credible, cheating would be deterred.

Consider this example in greater detail. To reduce the cost of abatement, it will pay countries to substantially invest in capital (energy conservation). Once the capital investment has been made, provided it is sunk, a country cannot credibly threaten to increase emissions substantially in the wake of a violation of the Convention by other countries – at least not quickly – because the cost of doing so would be prohibitive. However, such would not be the case if the country underinvested in capital to begin with. For then punishment is much cheaper; the saving in costs by not abating emissions is much greater when a country has invested in too little abatement capital. The effect of the strategic move is to make severe punishment credible. Notice that the strategic mover both loses and gains as a consequence of the move. In underinvesting in abatement capital, the country incurs higher abatement costs in every period. But in deterring cheating, the country gains from increased abatement by the countries that otherwise would have cheated. The strategic "investment" is worthwhile if this gain exceeds the associated loss.[9]

Depending on the nature of the game, it could also pay a strategic mover to overinvest in capital. The reason is that overinvestment would diminish a player's incentive to cheat. Consider the asymmetric prisoners' dilemma game depicted in Figure 3.2. If Country A can convince Country B that it will abate its emissions, then Country A will know that B will abate its emissions, and the full co-operative outcome will be achieved. One way country A can do this is by overinvesting in pollution control equipment. In doing so, the cost of abatement will fall, and hence the pay-off to abatement will rise and the pay-off to polluting will fall. Suppose the effect of the overinvestment is to increase the pay-off to abatement by one, and reduce the pay-off to polluting by one. Then we obtain the game shown in Figure 3.3. Abatement is now a dominant strategy for country A. Given that A will abate, B can do no better than to abate its own emissions.

It was shown earlier that binding commitments cannot be made because of the respect which international law accords sovereignty. However, international treaties can become part of the national law of signatories, and national law is harder to violate. Making treaties a part of national law can thus serve to weaken the ability of a country to cheat on the agreement. As an example, environmental groups have no standing in the International Court of Justice, but they do have standing in U.S. courts. An environmental group can take the U.S. government to court for failing to comply with a treaty. In Defenders of Wildlife, Inc. v. Endangered Species Scientific Authority, the U.S. courts prohibited the export of bobcat pelts until the government had complied with its obligations under the Convention on International Trade in Endangered Species (see Lyster, 1985, pp. 13-14). Of course, the U.S. government can eventually get out of its obligations by withdrawing from a treaty. But there is a certain advantage in being constrained by domestic law.

To see this, reconsider the symmetric prisoners' dilemma game shown in Figure 3.1. If both players are constrained by domestic law, the effect is to increase the penalty associated with cheating. Suppose the penalty to cheating is 2 points. Then we obtain the pay-off matrix shown in Figure 3.4. Abatement is now each country's dominant strategy.[10]

The global warming game is asymmetric; countries differ substantially. For example, treaties are in a sense more binding on some countries than on others. If the United States signs (and ratifies) a Convention, the threat of legal action by domestic environmental groups may make it much more costly for the U.S. to later renege on its commitment, even if reneging serves the purpose of punishing defectors. Depending on the nature of the game being played, it could be better for a country not to sign a Convention but to indicate its willingness to comply with the terms of the treaty. In this case, failure to sign the Convention makes punishment easier. (Interestingly, the Montreal Protocol does not impose punishments on nonsignatories that are in full compliance with the obligations of the Convention and have submitted data verifying full compliance. See Paragraph 8 of Article 4 of the revised Protocol.)

Figure 3.3. **Modified asymmetric prisoners' dilemma game**

		Country B: Abate	Country B: Pollute
Country A	Abate	6,5	3,4
	Pollute	5,2	2,3

Payoffs to (Country A, Country B)

Figure 3.4. **Modified symmetric prisoners' dilemma game**

		Country B: Abate	Country B: Pollute
Country A	Abate	5,5	2,4
	Pollute	4,2	1,1

Payoffs to (Country A, Country B)

Widening the game

The punishment mechanisms discussed earlier have the flaw that the punishing countries harm themselves as well as the "cheating" countries. This acts as a major impediment in making severe punishment credible.

One way of making severe punishment more credible is for the punishing countries to choose a different instrument – one that harms the country being punished much more than the one making the punishment. An example given in Barrett (1991*b*) is the Pelly Amendment to the U.S. Fisherman's Protective Act of 1967. This law allows the U.S. to punish countries that violate international fishery agreements or international agreements intended to protect endangered species by prohibiting imports of fish products from the offending country. It seems inevitable that negotiators to a global warming Convention will search for similar instruments, a possibility recently suggested by Whalley (1991, pp. 186-187):

> "In theory, it is conceivable that a global treaty on carbon emissions could specify a penalty system which would allow other countries to deviate from carbon emission reduction targets if particular countries did not meet their own commitments. What seems more likely, however, is that large countries would use threats of actions in other non-environmental policy areas to enforce environmental treaty commitments."

> "The instrument which is most commonly suggested for this purpose is trade threats."

This attempt to widen the game is similar to the approach suggested in the Chapter 2 discussion of side payments. There, it was suggested that introducing other issues to the global warming negotiations might enable agreement to be reached without the need for monetary side payments, or at least with a reduced need for such side payments. However, there is a difference between the two cases. Widening the game to facilitate side payments is generally helpful. Two issues are more effectively resolved in combination than if negotiated separately. Here, that outcome seems less likely. In using trade as a deterrent to free-riding in a global warming Convention, there is the prospect that any benefit to increased co-operation on global warming may be offset, at least partially, by losses resulting from barriers to trade. Threats to restrict trade would be beneficial if they never had to be implemented. But that outcome could not be assured, and in trying to win co-operation in a global warming Convention, negotiators may only succeed in undermining co-operation on trade policy.

CONCLUSIONS

As challenging as it will be to reach an international agreement limiting greenhouse gas emissions, *sustaining* such an agreement is likely to be even harder because of the strong incentives countries have to free-ride. These free-rider incentives essentially derive from the high costs of reducing greenhouse gas emissions substantially, and the fact that a great many countries would have to co-operate to reduce global emissions. Because of the number of countries involved, the contribution of any one country to global emission reductions is small. Hence, each country suffers only a small loss in benefit by reducing its abatement. But each country enjoys a substantial saving in costs, since the costs of abating substantial quantities of emissions are large. The net benefits to each country of reneging on a Convention limiting greenhouse gas emissions (or not signing such a Convention in the first place) are therefore substantial (at the margin).

Whether failure to deter free-riding will reduce global net benefits substantially depends on whether global benefits from abatement are large. One economic analysis suggests that these global benefits are not large. If this analysis is correct, then failure to deter free-riding will not cause great harm, because countries would not want to abate their emissions substantially even if co-operation were full. However, there is great concern that global benefits from abatement could be large.

The free-rider problem is made worse by the effect abatement by signatories will have on the costs (and not simply the benefits) of abatement by nonsignatories. If signatories reduce their emissions, world demand for fossil fuels will shift downwards, and hence equilibrium prices will fall. This will make it attractive for nonsignatories to increase their consumption of fossil fuels, and hence to increase their emissions.

To deter free-riding, a global warming Framework Convention must include mechanisms that punish countries that fail to comply with the terms of the Convention, or that fail to sign the treaty in the first place. Such punishment must be credible; if the mechanism is triggered, then it must be in the interest of the countries called on to make the punishment to fulfil that obligation.

One potential mechanism would link abatement by signatories (and, generally, countries in compliance with the terms of the Convention) to abatement by potential nonsignatories. This could be accomplished by ensuring that the general abatement obligations of signatories would not be implemented until a minimum number of countries representing a minimum share of global emissions had signed the Convention. Alternatively, the Convention could impose obligations on all signatories that varied with the number of signatories and their share of global emissions.

These mechanisms could be strengthened by procedures governing how these obligations could be achieved. If countries overinvest in abatement technology, and if this investment is sunk, then the incentive these countries have to "cheat" is reduced. Alternatively, if countries underinvest in abatement technology, then the incentive these countries have to punish cheaters is increased. The Framework Convention may therefore want countries with strong incentives to cheat to overinvest, and countries with important roles to play in enforcement to underinvest, in abatement technology.

It is unlikely that these mechanisms and procedures will eliminate free-riding. The great problem is that while they harm potential cheaters, they also harm the countries called upon to make the punishment. There will therefore exist strong incentives to search for other mechanisms – ones that harm cheaters much more than the countries making the punishment. The obvious example is that of trade restrictions. However, widening negotiations to include trade sanctions could undermine international negotiations on free trade, and the benefits that increased co-operation in this area could bring.

Notes

1. Preliminary analyses of both the science and the impacts of climate change suggest that some countries may actually benefit from global warming, at least up to a doubling in carbon dioxide concentrations. However, this Chapter assumes that *all* countries would benefit from reductions in greenhouse gas emissions. Hence, it also assumes that all countries would prefer that climate change be limited.

2. "Substantially" should be taken to mean a level at which the marginal cost of abatement for any country equals the global marginal benefit of abatement. Nordhaus (1990) argues that the efficient abatement level is much smaller than the targets being proposed by certain scientists and politicians.

3. A difficulty with this equilibrium concept is that the assumed behavioural responses are contradicted away from the equilibrium. The Nash assumption is that each country expects other countries *not* to respond, if it alters its own abatement levels. And yet, other countries *do* respond. An alternative approach is to employ a "consistent conjectures equilibrium" assumption. In the neighbourhood of such an equilibrium, the actual responses by countries conform to the assumed responses. Cornes and Sandler (1983) show that this equilibrium concept implies that free-riding will be even more of a problem using this assumption.

4. This assumes that countries are identical.

5. Note that, while it may be desirable to ban the export of technologies that can reduce the damage from potential climate change, technologies that can reduce abatement costs should not be banned. The availability of such technologies will lower the costs to non-signatories of acceding to the Convention, and increase the abatement effort in countries that still find it desirable to remain outside the Convention.

6. For the parties that want to stick to this agreement, it must be the case that the game is played over an infinite period of time, or that there always remains a positive probability that the players will meet again. To see this, consider an example. Suppose that two players know they will meet ten times, that player 1 would like to take player 2 to arbitration the first time, that player 2 would like to take player 1 to arbitration the second time, and so on, until the tenth time they meet, when player 2 would like to take player 1 to arbitration. Because the players will never meet an eleventh time, player 1 can credibly refuse to go arbitration the tenth time. Knowing that player 1 can credibly refuse to go to arbitration the tenth time, player 2 can credibly refuse to go to arbitration the ninth time. And so on. The requirement that players always meet again, or have a positive probability of meeting again, is very reasonable in the case of global warming.

7. Barrett's ((1991*a*) model predicts that both factors would have an influence on strengthening the agreement.

8. Guttman's (1978) analysis is based on the provision of public goods, where the choice variables are "flat" and "matching" rates for donations to provide the public good. Abatement of greenhouse gases is a global public good, but the choice variables are abatement levels, rather than donation levels.

9. This strategy of underinvesting is called the "lean and hungry look". For a discussion of strategic behaviour in oligopolies, see Shapir (1989).

10. Analogous strategic behaviour in oligopolies include marketing practices. For example, a price protection clause, which guarantees past customers that they will be refunded if price is later reduced, serves to make the "drop the price" strategy more costly. See Scherer and Ross (1990), Chapter 6.

References

BARRETT, S. (1990). "The Problem of Global Environmental Protection," *Oxford Review of Economic Policy,* 6: 68-79.

———, S. (1991a). *The Paradox of International Environmental Agreements,* mimeo, London Business School.

———, S. (1991b). "Economic Analysis of International Environmental Agreements," in OECD, *Responding to Climate Change: Selected Economic Issues,* Paris: OECD.

BIRNIE, P. (1988). "The Role of International Law in Solving Certain Environmental Conflicts," in J.E. Carroll (ed.). *International Environmental Diplomacy: The Management and Resolution of Transfrontier Environmental Problems,* Cambridge: Cambridge University Press.

BOHM, P. (1990). *Mitigating Effects on Fuel Prices from Incomplete International Cooperation to Reduce CO_2 Emissions,* mimeo, Department of Economics, Stockholm University.

CORNES, R. and T. Sandler (1983). "On Commons and Tragedies," *American Economic Review,* 73: 787-792.

FISCHER, W., J.C. di Primio, and G. Stein (1990). *A Convention on Greenhouse Gases: Towards the Design of a Verification System,* Jülich: Forschungszentrum Jülich GmbH.

GUTTMAN, J.M. (1978). "Understanding Collective Action: Matching Behavior," *American Economic Review,* 68: 251-255.

LYSTER, S. (1985). *International Wildlife Law,* Cambridge: Grotius.

NORDHAUS, W.D. (1990). *To Slow or Not to Slow: The Economics of the Greenhouse Effect,* mimeo, Department of Economics, Yale University.

SCHERER, F.M. and D. Ross (1990). *Industrial Market Structure and Economic Performance,* Boston: Houghton Mifflin.

SHAPIRO, C. (1989). "Theories of Oligopoly Behavior," in R. Schmalensee and R.D. Willig (eds.). *Handbook of Industrial Organization,* Volume 1, Amsterdam: Elsevier.

SPRINGER, A.L. (1988). "United States Environmental Policy and International Law: Stockholm Principle 21 Revisited," in J.E. Carroll (ed.). *International Environmental Diplomacy: The Management and Resolution of Transfrontier Environmental Problems,* Cambridge: Cambridge University Press.

TASK FORCE ON THE COMPREHENSIVE APPROACH TO CLIMATE CHANGE (1991). *A Comprehensive Approach to Addressing Potential Climate Change,* U.S. Department of Justice, Washington, D.C., February.

TRAIL SMELTER ARBITRAL TRIBUNAL (1941). "Decision," *The American Journal of International Law,* 35: 684-736.

WHALLEY, J. (1991). "The Interface Between Environmental and Trade Policies," *Economic Journal,* 101: 180-189.

MAIN SALES OUTLETS OF OECD PUBLICATIONS – PRINCIPAUX POINTS DE VENTE DES PUBLICATIONS DE L'OCDE

OECD PUBLICATIONS, 2 rue André-Pascal, 75775 PARIS CEDEX 16
PRINTED IN FRANCE
(97 92 07 1) ISBN 92-64-13668-1 - No. 46017 1992